# CHESAPEAKE WATERS
Pollution, Public Health, & Public Opinion, 1607–1972

# CHESAPEAKE WATERS

*Pollution, Public Health,
and Public Opinion, 1607-1972*

by John Capper,
Garrett Power,
and Frank R. Shivers, Jr.

Tidewater Publishers
Centreville, Maryland

Copyright © 1983 by Tidewater Publishers

Library of Congress Cataloging in Publication Data

Capper, John R., 1937-
    Chesapeake waters.

    Bibliography: p.
    Includes index.
    1. Water—Pollution—Hygienic aspects—Chesapeake Bay Region (Md. and Va.)—History. 2. Marine pollution—Hygienic aspects—Chesapeake Bay (Md. and Va.)—History. 3. Water—Pollution—Environmental aspects—Chesapeake Bay Region (Md. and Va.)—History. 4. Marine pollution—Environmental aspects—Chesapeake Bay (Md. and Va.)—History. 5. Water—Pollution—Chesapeake Bay Region (Md. and Va.)—Public opinion—History. 6. Marine pollution—Chesapake Bay (Md. and Va.)—Public opinion—History. 7. Environmental policy—Chesapeake Bay Region (Md. and Va.)—History. 8. Chesapeake Bay Region (Md. and Va.)—History. I. Power, Garrett. II. Shivers, Frank R., 1924- . III. Title. [DNLM: 1. Water pollution—Virginia. 2. Water pollution—Maryland. 3. Public health—Virginia. 4. Public Health—Maryland. 5. Public opinion. WA 689 C248c]
RA592.C45C36 1983 363.7'394'0975518 83-40102
ISBN 0-87033-310-0

Manufactured in the United States of America

First Edition

*This book is for our children*
Rachel and Nathan Capper
Kate, Amy, and Nellie Power
Maggie, Natalie, Lottchen, and Philip Shivers

# CONTENTS

# LIST OF ILLUSTRATIONS

# PREFACE

The Chesapeake Bay is the most studied and best understood estuary in the United States. Yet, it is practically unexamined in the areas of the social sciences and the humanities. While millions of dollars have been spent on producing the thousands of studies that examine the physical, biological, chemical, and engineering aspects of the Bay, little attention has been given to understanding the political, cultural, and economic character of Bay governance.

The relationship of the governments of Maryland and Virginia to the Bay is imperfectly documented. Government documents which do exist are scattered in various libraries in both states and have not found their way into the numerous bibliographies that have been assembled for the Bay. In Virginia, the State Water Control Board did not produce annual reports until 1972, the cutoff date for this study. In Maryland, the reports of water-quality agencies tend to be perfunctory and repetitive, and they give little indication of the real issues facing the agencies over the years. The many planning documents which do exist (the recent Corps of Engineers' Chesapeake Bay Study is the largest) are general compilations of information and issues rather than original pieces of research.

As a result, the present study has had the benefit of little scholarship to point the way. The researcher is forced to approach his material as though he were an archeologist, finding a few shards here, a few bone fragments there. Piecing together a coherent story out of the fragments requires a certain amount of logic, a workable hypothesis about the overall nature of the creature to be described, and some theories about how the evidence fits together.

But the story is worth the telling. After all, the quality of Chesapeake Bay is a matter of public opinion as well as scientific opinion. Those concerned about the Bay must understand the human-political dimension as well as the physical-biological side.

We relied primarily on written sources. Those proving most fruit-ful have been the annual reports of various state agencies, the occa-sional reports of study commissions and blue ribbon panels, and the codes, statutes, and case law of the two states. Agency files proved difficult to use because they are boxed and stored, full of irrelevant material, unorganized, and uncataloged.

Interviews with persons familiar with Bay issues have given a general orientation to a particular period and suggestions of topics of sources for further research. We have not attempted to get detailed information of specific events through such interviews. The written record, we feel, stands on its own.

In particular, we also made use of the abundant collections of newspaper files in libraries. While newspaper articles may have ques-tionable accuracy, they identify key issues and place them defini-tively in time. Without them, numerous controversies, left only to the official archivists, would go unrecorded. In this study, informa-tion from newspapers gives a sample of issues and shows the broad trends in water-quality awareness. Feature articles in magazines and newspapers are particularly useful, because they both reflect, and partially shape, the public attitudes toward the Bay. Changes in these attitudes provide data used throughout the report.

We hope that this book has something to say that has been neglected in the public debate over the Bay. Its conclusions do not mean that scientists should be involved less in research on the Bay. They simply suggest that economists, political scientists, historians, and lawyers, should be involved more.

# ACKNOWLEDGMENTS

The authors thank Randall Beirne, William A. Cook, Marian Dillon, Richard McLean, David P. Miller, Sandralee P. Morris, Erik Olson, Natalie W. Shivers, LuAnn Young, H. Chace Davis, Jr., and Scott Lewis Zeger, all of whom made special contributions. We are likewise indebted to Penelope R. Power who prepared the index. The authors are also grateful to Gilbert Byron and his publisher, The Driftwind Press, for permission to quote from "Tangier Prayer" (1942), to Annie Dillard and her publisher, Harper's Magazine Press, for permission to quote from *Pilgrim at Tinker Creek* (1972), and to Alfred A. Knopf, Inc. for permission to quote H. L. Mencken from *Newspaper Days* (1941) and *Happy Days* (1940).

Preparation of the initial report upon which this book is based was funded in part by the Chesapeake Bay Program of the United States Environmental Protection Agency (EPA Grant X-003226-01). It has been approved for publication. Approval does not signify that the contents necessarily reflect the views and policies of the United States Environmental Protection Agency, nor does mention of trade names or commercial products constitute endorsement or recommendation for use.

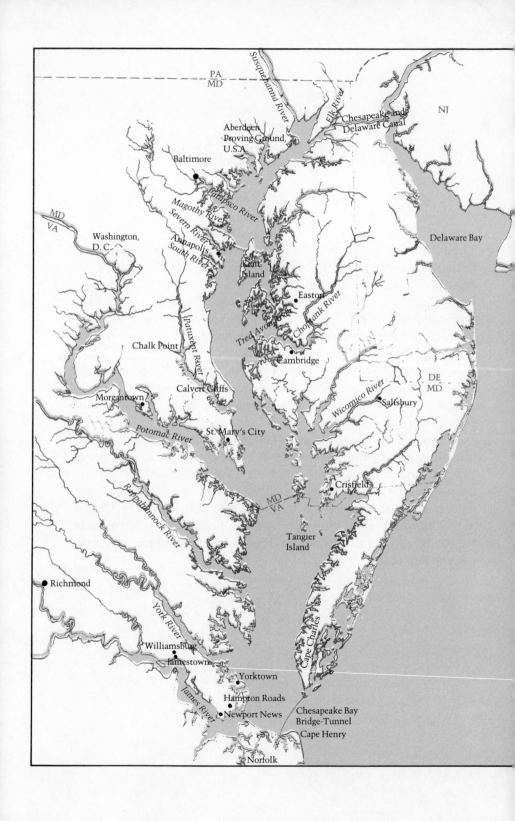

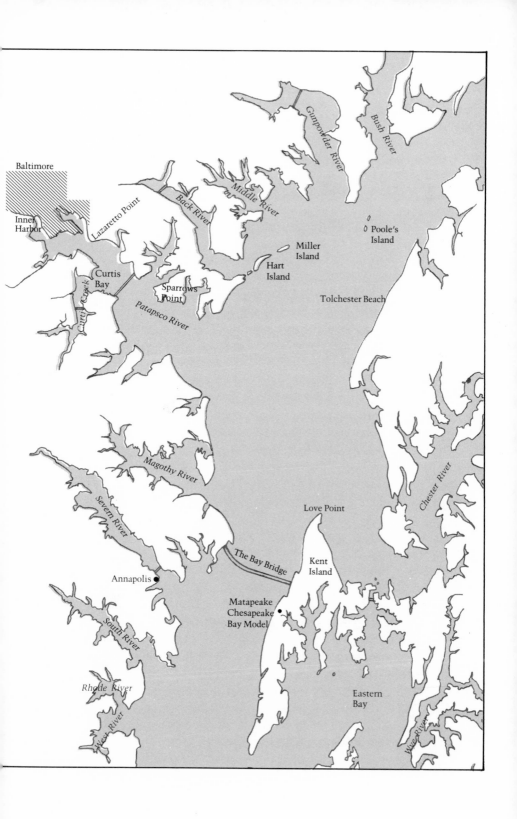

Baltimore

Inner
Harbor

Curtis Creek

Curtis
Bay

Lazaretto Point

Sparrows
Point

*Patapsco River*

Back River

*Middle River*

*Gunpowder River*

Bush River

Poole's
Island

Miller
Island

Hart
Island

Tolchester Beach

*Magothy River*

*Severn River*

Love Point

Chester River

*The Bay Bridge*

Kent
Island

Annapolis ●

Matapeake
Chesapeake ●
Bay Model

*South River*

*Rhode River*

Eastern
Bay

*West River*

*Wye River*

# CHESAPEAKE WATERS
Pollution, Public Health, & Public Opinion, 1607–1972

Landing at Jamestown. Engraved for the *New York Mirror*, May 8, 1841. This nineteenth century romanticized version of the first permanent settlement by Europeans north of Spanish America is reminiscent of the exaggerated idealization of settlement prospects reflected in much of the writing of the early seventeenth century about the New World. Courtesy the Library of Congress.

# From Jamestown to Tropical Storm Agnes

There were never Englishmen left in a forreigne Country
in such miserie as we were in this new discovered Virginia.
... Our drinke cold water taken out of the River, which was
at a floud verie salt, at a low tide full of slime and filth,
which was the destruction of many of our men.
—George Percy (1625)

During 1972 Tropical Storm Agnes released record
amounts of rainfall on the watersheds of most of the major
tributaries of Chesapeake Bay. The resulting floods, cate-
gorized as a once in 100 to 200 year occurrence, caused
perturbations of the environment in Chesapeake Bay, the
Nation's greatest estuary.
—Chesapeake Research Consortium, *The
Effects of Tropical Storm Agnes on the
Chesapeake Bay System* (1975)

In 1607, the first English colonists in the Chesapeake Bay region drew
their drinking water from the often brackish James River estuary,
which ebbed and flowed past the shores of the swampy island that
was to be called Jamestown. In doing so, they exposed themselves to
disease-carrying bacteria from their own wastes (according to the
theory of a twentieth-century historian) and subjected themselves to
gradual salt poisoning from the brackish water.[1]

Three hundred and sixty-five years later, in 1972, tropical storm
Agnes swept through the region, dumping record-breaking quantities
of rain throughout the sixty-five thousand square mile basin that
drains into the Bay. Its force was described by an observer, the writer
Annie Dillard:

It's all I can do to stand. I feel dizzy, drawn, mauled. Below me
the floodwater roils to a violent froth that looks like dirty lace, a
lace that continuously explodes before my eyes. If I look away,
the earth moves backwards, rises and swells, from the fixing of
my eyes at one spot against the motion of the flood. All the
familiar land looks as though it were not solid and real at all, but
painted on a scroll like a backdrop, and that unrolled scroll has
been shaken, so the earth sways and the air roars.

Everything imaginable is zipping by, almost too fast to see. If
I stand on the bridge and look downstream, I get dizzy, but if I
look upstream, I feel as though I am looking up the business end
of an avalanche.[2]

The resulting floods stripped wastes from throughout the region and
dumped them into the Bay, leaving it a vast mess-roil, awash with
bacteria, loaded with massive quantities of organic matter, clogged
with debris ranging from plastic bottles to houses, and freshened past
the level of tolerance of saltwater-demanding plants and animals. In
every sense of the word, the Bay was grossly polluted. It had become
foul, noxious, dirty, and corrupted, the worst condition of its recorded
history.[3]

These two conditions, that of 1607, subtle and speculative, and
the one of 1972, blatant and documented, serve as an appropriate
beginning and end for this study, a historical survey of water quality
issues on Chesapeake Bay. They bracket the period of this study—
from 1607 and European settlement to 1972 and modern megapolis.
They reflect the range of issues from the microscopic and localized, to
the obvious and Baywide. They illustrate that water quality, although affected by human activity, is to a large degree independent of
that activity. And they reflect that conditions of water quality may
require sophisticated tools of science to detect what may be obvious
to the untrained observer.

Between these two events, the people of the region have taken a
great interest in the Chesapeake Bay. This book traces the development and evolution of their governments' attention. Focus is on
human institutions, ideas, and attitudes, rather than on the physical,
chemical, and biological attributes. We examine: how the ideas about
water quality have been formed, changed, and communicated; how
government has organized itself to the task of studying and influenc-
ing the quality of the Bay; and how perceptions of the Bay have
brought about change in the way it is used.

Our thesis comes from the view that, while it is important to describe the conditions of the Bay as accurately as human science and technology allow, it is just as important to understand how the concept of water quality has operated in the political process. This book, then, is an attempt to broaden the debate about the Bay and its future and to include an understanding about how we and our predecessors have shaped and used our governments to have an influence on the quality of Bay water.

Now for definition of terms: *government* is broadly interpreted to include any activity or viewpoint sanctioned by government bodies. It includes actions by courts, legislatures, and administrative agencies—local, state, and national. During the nearly four hundred years' recorded history of the Bay, there have been shifts in the locus of governmental power. In the seventeenth and eighteenth centuries, the courts were the dominant law-making bodies. By resolving a number of *ad hoc* conflicts, they created what came to be viewed as a body of common law. Then, during the nineteenth century, the Maryland and Virginia legislatures attempted to address some problems of water quality. They enacted laws designed to resolve specific conflicts over fisheries and navigation.

Early in the twentieth century, talk grew of managing water quality by administrative agencies. Management may be a misnomer. Agencies do not manage the Bay in any generally accepted sense of the word. Although human influence on the Bay has been considerable (both intentional and unintentional), the Bay is still to a large extent beyond the ability of our technology to regulate or control. Nonetheless, an "administrative state" has evolved in which public administrators influence the Bay to a greater extent than courts or legislatures.

Our definition of *Chesapeake Bay* is expansive. In a physical sense it is to include the whole estuary—the semienclosed coastal body of water that has a free connection with the open sea and within which seawater is measurably diluted by fresh water derived from land drainage. So defined, the Bay includes long stretches of the Potomac, Rappahannock, York, and James, as well as hundreds of other rivers, creeks, bays, and sounds. The nomenclature of the Bay region is confusing. Some rivers are not fresh and flowing but subestuaries (for example, the Severn River); other rivers flow to the fall line and then become estuarine (for example, the Potomac River). Likewise, some creeks are tidal and others not. The words "bay" and

Two pencil sketches by John H. B. Latrobe illustrate uses of the upper Chesapeake about 1830. *Above,* commercial ships at the mouth of the Susquehanna River, Turkey Point, Maryland. *Below,* a raft on the Elk River, Maryland, on its way to a Chesapeake port. These calm waterscapes contrast with the flood of fresh water of Tropical Storm Agnes (1972) which so damaged the Bay. The Susquehanna River, even when not in flood, is the Bay's greatest single source of fresh water. Courtesy the Library of Congress.

"sound" are also loosely applied. Pocomoke Sound fits the c tional definition of a bay while Chincoteague Bay behind the beach on the Atlantic is perhaps more appropriately a sound study avoids semantic confusion by looking to the estuary regardless of the popular name.[4]

This study also looks beyond the estuary. Because the quality of the Bay is the result of all physical additions, we consider activities in the entire Chesapeake Bay watershed, which includes six states and reaches as far from the Bay as mid-New York State. The six states are Delaware, Maryland, New York, Pennsylvania, Virginia, and West Virginia; to these add the District of Columbia. The chief boundary dispute around the Bay, between Maryland and Virginia, still causes trouble between watermen.

A historical perspective does give meaning to the concept of water quality in Chesapeake Bay. For roughly the first two hundred years covered by this book, the only water-quality issues were siltation, the condition of the fishery, and public health. By the 1820s concerns were growing about the declining stocks of shad and herring. These declines were attributed in part to the effects of siltation, sedimentation, temperative changes due to land clearing, and the mechanical action of ships' wakes, as well as the blockage of streams by dams and raceways.

By the middle of the nineteenth century, people also recognized that fish avoided areas that were fouled by wastes, such as sawdust, canning wastes, and discharges from slaughterhouses and tanneries. The present usage of the word "pollution" then came about. Toward the end of the century, fisheries scientists were making the first formal studies of the relation of fish life cycles and productivity to environmental factors, both natural and man caused. By the twentieth century, the germ theory of diseases had added the concept of waterborne disease bacteria to the growing science of water quality of the Bay.

Once it was established that oysters were carriers of diseases originating from human sewage, oyster sanitation then became the most influential water quality concern. This connection was made in the 1890s, but had its first impact on the Bay shortly after 1900. Also about this time fishermen began complaining about the effects of wastes from food-processing plants. Thus began the concern for industrial wastes that for the first half of the century continued to come primarily from food-processing industries and other small producers of organic wastes.

By the end of World War I, oil from ships using the Bay was seen by fisheries officials and bathing beach operators as a major water-quality problem, and in the mid-1920s, oyster sanitation again became a major concern when several epidemics blamed on bad oysters brought on a wide scale closure of bars. Growth of municipal sewerage and industry, by the end of World War II, created conditions that made pollution in both tidal and nontidal streams a major concern.

In the 1950s, scientists began to investigate more subtle water-quality factors such as the effects of insecticides, fertilizers, and heavy metals on marine organisms. That investigation ushered in the era of concern for exotic discharges to the system, which remains a predominant concern today. By about 1960, a new set of concerns was added to the list, including thermal discharges, overenrichment, general land runoff, and freshwater diversions. Also in the 1960s, the concerns expanded from those of specific conditions and pollutants to a general worry about the overall ecological health of the Bay—the concept of water quality shifted from pollution to ecology.

This study takes a historical look at the quality of Chesapeake Bay waters and the measures that governments have taken to influence that quality. Because most conflicts involve the issue of water quality, this book in some respects is a history of Bay uses in general. The organizing principle and point of view, however, stress the relation of various uses to water quality, and the way in which government has intervened to have some effect on that quality.

The book's organization is both chronological and thematic. We describe the nature of the Bay in Chapter 2, and the uses and abuses of Chesapeake waters during the first three hundred years of European settlement in chapters 3, 4, 5, and 6. In chapters 7, 8, 9, 10, and 11 we take up the major conflicts that have arisen with respect to the Bay in the twentieth century, in more or less chronological order. In Chapter 12 we state our conclusions.

# TWO

# Noble Arm of the Sea

The Chesapeake Bay is a noble arm of the sea.
—Lord Morpeth (1842)

Virtually all written documents about Chesapeake Bay, whether technical or popular, begin with a description of the Bay and the region. Because these descriptions contribute to the composite public perception of the Bay, we would like to make some general observations about these Bay descriptions before presenting our own.

## DESCRIBING THE BAY

In general, descriptions of the Bay show the lack of a single authoritative source of information about the Bay, its resources, and the region that surrounds it. Those descriptions that do exist are sometimes inaccurate and present problems of definition and interpretation. Several examples will illustrate these problems.

One might expect that physical dimensions of the Bay, which are for the most part constant and accessible to measurement, would be readily available and accurate. Such is not the case. Anyone examining the literature of the Bay, whether recent or old, technical or popular, will find wide discrepancies in the various dimensions. Length, for example, is one such dimension. One would expect to find some variation because there is some judgment involved in such matters as placing the starting and ending points. Where does the Bay begin and the ocean end, for the mouth; and where does the Bay begin and the Susquehanna end, for the head. Another question is how to deal with the curves of the Bay. Nonetheless, these considerations do not account for the wide variation.

In fact, the length of the Bay has been authoritatively reported to be approximately 190 miles, using a method of measurement de-

Captain John Smith's map of Virginia (1627), the chief source of information on the New World for sixty years. Courtesy the Library of Congress.

scribed in a Chesapeake Bay Institute special report.[1] We came very close to duplicating this figure using a 1:200,000 scale chart and a map wheel. Yet, in the Corps of Engineers massive Chesapeake Bay study, the Bay is reported to be over 200 miles long,[2] and in 1979, two widely published Bay authorities coauthored a paper in which the length is listed as 168 miles (271 kilometers), which is even less than the straight-line distance between the lines of latitude of the mouth and head of the Bay (37 degrees north to 39 degrees 32 minutes, or 152 nautical miles = 175 statute miles).[3]

Hence, while there are sources of accurate information, it is difficult for all but the most diligent reader to sort out the good from the bad. Because many writers of descriptions draw on previously published sources without question of citation, old errors are perpetuated.

Problems of definition complicate efforts to relate census data to the Bay. Descriptions which give population figures, and at least imply that there is a direct relation between population and the stresses of pressures put on the Bay, pose two questions. First, there is always a question of how to define the Bay region. A description might range from including a narrow strip of land around the Bay to including the entire watershed. Second, and more perplexing, is the question of the relation of human population to the Bay. Impact on the Bay is going to differ, depending on location, method of sewage disposal, predominant land use, and employment. It might be that five thousand watchmakers living in metropolitan apartments would have less impact on the Bay than one dairy farmer living on the shores of an oyster-producing creek. As a result population figures may then lack the analysis necessary to make them relevant to Bay issues.

Perhaps no numbers signify more for Bay water-quality issues than fisheries statistics. Much of the political energy that drives the Bay system is generated over public concern for the state of the fishery. Problems associated with the collection of these numbers are legion. They include: methods of collecting information have changed markedly over the years; methods vary between the states; sport catches are unknown; harvests are affected by economic and cultural forces (low prices and war, for example) having nothing to do with the size of the fish population; fishermen have been suspected of underreporting catch to avoid taxes. Moreover, fluctuations in fish populations, and, therefore, catch, are subject to wide variations due to unknown or poorly understood factors. These factors could be natural or human in origin.

Serious study has gone into the examination and analysis of fisheries statistics, and there is continuing advance in their utility as management tools.[4] But over the historical period covered by this book, these statistics have tended to be used uncritically, particularly when inferences are drawn from them about the effect of man's activities on the water quality of the Bay.

## CHARACTERISTICS OF THE BAY

Forewarned by our own criticism, we now undertake our description of Chesapeake Bay. It is intended to be relevant to the specific topic of study contained in this book—man's effect on water quality over time.

### Geographic Characteristics

Common practice compares Chesapeake Bay to other large estuaries or lakes. One of its striking features is that it has a very long shoreline in relation to its overall length or size. Taking the previously discussed length figure of one hundred and ninety miles and a currently prevalent figure for shoreline of about eight thousand miles, the Bay has a shoreline to length ratio of over forty to one. Lake Ontario, the smallest of the Great Lakes, has a ratio of about two and a half to one. For Chesapeake Bay, there are roughly two miles of shoreline for every square mile of surface area; for Ontario, there are about eighteen square miles for each mile of shoreline.

These ratios point to at least three important aspects of the Bay. First, they are another way of expressing the extreme branchiness of the Bay: the Bay is divided into a large number of separate components, the seemingly numberless bays, creeks, and guts that so frustrated colonial land travellers, but so fascinates the contemporary "gunkholer." Second, the extensive shoreline allows for a high level of human access to the Bay, both to use and to enjoy, and to affect it by land activities. Third, the land-water edge is an area of relatively high biological activity, and probably contributes materially to the productivity and biological variety of the Bay.

Another striking characteristic is its shallowness. This is apparent to the visitor of the Bay hydrologic model in Matapeake, Maryland. There, at a horizontal scale of 1,000 to 1, is a concrete Bay in miniature. The Bay itself is over fifteen hundred feet long. Since it is enclosed under a single roof, there is a sense of vastness about the

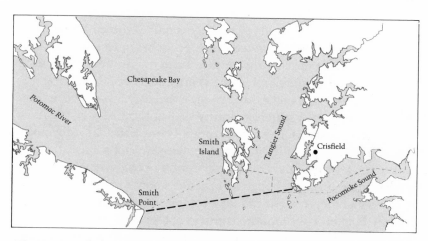

The section of Chesapeake Bay showing the shifting boundary between Maryland and Virginia, long a matter of conflict. The dark line is the boundary of 1668 and the light line is the present boundary

place. Yet, the deepest part of the Bay is only eighteen inches deep, and most of the area underwater is three inches or less deep. It must then be remembered that the vertical scale is only 1 to 100. Thus, if the model were at the same scale in both directions, the deepest part would be about two inches, and most of the "Bay" would be covered by a film of water scarcely worthy of being called wet. Again, a comparison to Lake Ontario is in order. There, a similar scale model would have a depth of eight feet at its deepest, and a large area over five feet. It would make a fine swimming pool, whereas the Bay model isn't even deep enough for a toddler to wade in. The Bay proper, with its eighteen trillion gallons of water, is a veritable drop in the oceanic bucket. Lake Ontario contains approximately forty times as much water. If dilution is the solution for pollution, Chesapeake Bay has a relatively limited capacity.

Finally, it is popular to speak of the Bay as a single system. Yet, in a profoundly important way, it is divisible into many separate sub-systems and in a way that many other large bodies of water are not. The rate at which water is exchanged or flushed from a subestuary to the estuary is often small or negligible. One effect of this geographic feature has been to make local conditions more acute than they might otherwise have been at the point of discharge. For example, a sewage treatment plant in Baltimore, Maryland, has fundamentally altered the quality of Back River on which it is located. On the other hand, effects from these wastes are more isolated and contained than they would have been if discharged into the main stem of the Bay.

Demographic Characteristics

We have already spoken of the difficulty of relating raw population figures in any direct way to the condition of the Bay. However, several potentially important generalizations are possible. First, municipal-industrial influences, whatever they might be, are concentrated in a relatively small area in and around the cities of Baltimore, Washington, Richmond, and Norfolk. The remainder of the Bay region is relatively lightly populated. Second, these areas have all experienced rapid growth since World War II, and will probably continue to do so. Land use changes within the standard metropolitan statistical areas are substantially from forested and agricultural to residential and commercial, with attendant effects on stream runoff and waste loads.

Third, the region as a whole has substantially less heavy industry than comparable regions in the East or Midwest.[5] Employment in government and service sectors, including military, is higher than the regional average. Fourth, even though the region has the image of being an important seafood producer, and fishing provides an important source of employment in some of the smaller tidewater counties of both states, from a regional standpoint the contribution of the fisheries industry to the economy is almost negligible.[6]

### Physical-Chemical Characteristics

Along with demographic characteristics, we must understand the major physical and chemical features of the Bay. These have been succinctly and clearly stated by Dr. Donald Pritchard.[7] Several important aspects might be usefully highlighted for the layman. First, waters in the Bay and its principal tributaries are in constant motion, gravity acting differentially on the denser salt water being overriden by the lighter fresh water entering from the rivers. This two-layered wedge effect accounts for the presence of seawater (much diluted) near the heads of the Bay and its tributaries, and the eventual exit of river water to the sea. This basic circulation is much affected by variations in river flow, and combinations of winds and atmospheric pressure changes that can drive very large quantities of water into or out of the Bay.

Second, the circulation of many of the smaller arms of the Bay are relatively sluggish because they do not have a substantial freshwater inflow to establish their own flow system. Their circulation is much affected by the differences in salinities between their lower reaches and that of the Bay. At certain times of the year they can undergo a rapid flushing because of strong differences between their salinity and that of the Bay. At other times there may be very little exchange. In some respects, then, for substantial periods of the year, these smaller bodies of water are essentially separate from the Bay. This condition applies as well to the smaller branches of the principal tributaries.

Third, the fate of a constituent added to the waters of the Bay is often a complex one. Silt from land erosion may interact, for example, with bacteria from a farm feedlot; nutrients from a sewage treatment plant might combine chemically with the wastes from an industry. A toxic chemical might be buried harmlessly in bottom sediments, or it might be physically and biologically transported throughout the Bay.

Fourth, alterations made by man can have a considerable impact on the physical and chemical oceanography of the Bay. A system of dams might substantially alter the flow regimen of a river entering the Bay. A navigation channel might affect the exchange rate between a tributary and the main Bay. Industrial and municipal wastes, moreover, can greatly increase the concentrations of naturally occurring substances, or add new and potentially harmful ones.

On the other hand, the natural forces affecting the Bay, and their range of variation, are enormous when compared in general to the scale of effects that are influenced or controlled by man. It is important to keep these forces in perspective when reviewing the history of man's concern for his impact on the quality of the Bay.

### Biological Characteristics

A fitting companion to Dr. Pritchard's paper just cited, and bound next to it in the report of the 1968 Governor's Conference on the Chesapeake Bay, is Dr. L. Eugene Cronin's description of the biology of the Bay.[8] First, as the Bay is physically and chemically a highly variable environment, so are the biological populations that fill the various Bay niches. Populations and reproductive success of numerous important species, including the oyster, the crab, and the striped bass, are subject to cycles that are as yet only partly understood. Understanding of these natural variations is confounded by the problem of assessing the influence of man, either through his harvests or through his indirect effect on water quality.

Second, the animals of the Bay combine permanent residents, regular and Bay-dependent visitors, occasional visitors, sometimes in great abundance, and strays. The biological openness of the system is one of its prominent features.

Third, the biology of the Bay poses a paradox. The Bay combines toughness and resiliency to change, on the one hand, and fragility and vulnerability, on the other—the former because the residents of the Bay are adapted to the great natural variability, and the latter because of the relatively short food chains in the Bay.

This nonnumerical description of the characteristics of the Bay has stressed openness, variability, and complexity. These concepts serve as prelude to our historical study. The Bay is also open in a political and economic sense. Man's use of the Bay is very much affected by events taking place outside the region. The recent buildup of demand for coal-shipping facilities, a result of the world energy market, is a case in point. The ports of Baltimore and Norfolk funnel coal to the world as they have for 150 years. Through this funnel of a bay pour products from the hinterlands of Virginia, Maryland, West Virginia, Pennsylvania, and parts of New York and North Carolina. Today these two Bay ports are among the top five in the United States for volume of trade, surpassed only by New York on the East Coast. That fact must be kept in mind as we study Chesapeake history.

We learn the importance of facts when we realize how the variability of the Bay's quality produces popular misunderstandings. A large fish kill (or, more appropriately, a fish die-off), is typically taken as prima facie evidence of pollution. To argue without other evidence that there is a direct causal link, however, is to commit the fallacy of post hoc, ergo propter hoc (after this, therefore because of this). This line of argument, regardless of what some ancient Greek logicians might think of it, has been often employed on the Chesapeake.

# THREE

# Many Deep-Water Ports

It is called the Bay of the Mother God, and in it there are
many deep-water ports, each better than the next.
—Brother Carrera (1572)

This first European record of the Chesapeake also described the Bay
as great and beautiful.[1] The year was 1572, and the writer was a
Spanish priest, Brother Carrera. But it was the gleam of commerce in
Carrera's eye that produced fateful consequences. Newcomers to the
New World arrogantly claimed it by "right of discovery." In their
minds bloomed the image of wealth, of riches greater than Spain's in
Latin America. The chief exploiter of this dream, Sir Walter Raleigh,
hired Richard Hakluyt to write *Discourse on Western Planning* in
order to convince Queen Elizabeth and her counsellors to support
settlements on a large scale and with a lot of royal backing. This work
pointed out advantages to the prosperity of the country:

> By makinge of shippes and by preparinge of thinges for the same,
> by makinge of cables and cordage, by planting of vines and olive
> trie, and by making of wyne and oyle, by husbandrie, and by
> thousandes set on worke.[2]

This visionary book imagined the towns "upon the mouthes of
the greate navigable rivers." And Raleigh then proposed his second
settlement in the New World to be established on Chesapeake Bay.
On January 7, 1587 it was incorporated by the governor and the
"Assistants of the City" of Raleigh, Virginia. Raleigh's propaganda
and overtures inspired enthusiasm in others, although his own plans
came to naught.

The next move towards the Bay was better planned. The Council
for Virginia of the London Company spelled out the moves for the
new colonists. Among the instructions, one provided that colonists

A late sixteenth century scene painted by John White. Hariot, the engraver, noted that the Indians have many fish of kinds never found in Europe, and all of an excellent taste. Of the Indians, he said, "They are untroubled by the desire to pile up riches for their children . . . sharing all those things with which God has so bountifully provided them." Courtesy the Library of Congress.

must not "plant in a low or moist place because it will prove un-
healthful." This advice, unfortunately, the colonists disobeyed when
they chose Jamestown as a site.

Much has been written about the unhappy years after the choice
of the hundred settlers in the spring of 1607. One historian has
conjectured that the colonists poisoned themselves with salt from
the brackish waters and infected themselves with pathogens from
their own wastes.[3] This detective work in public health may or may
not be accurate, but it is certainly the case that by October of 1608
over one-half of the two hundred-odd newcomers had died.

Not all experiences were proved bad. In April 1607, one colonist
reported of Lynnhaven Bay that "we got good store of mussels and
oysters, which lay on the ground as thick as stones. We opened some
and found in many of them pearls." That same month at Cape Henry
colonists broke up an Indian oyster roast and pulled from the fire
oysters "very large and delicate in taste." Two months later a letter
from "Council in Virginia to Council in England" told of a Virginia
river notable for "sweetness of water" and added that it was "so
stored with sturgeon and other sweet fish as no man's fortune has
ever possessed the like."[4]

After the first winter at Jamestown, Captain John Smith led a few
men in a shallop on a voyage around Chesapeake Bay. The Bay proved
to be a monster. Its winds, rains, mosquitoes, and Indians tested the
man famous for bravery, boastfulness, and energy. He mapped the
Eastern Shore as best he could. But the complex interweaving of
water and land defied him, observant though he was. Smith Island
recalls his visit in its name.

On the western shore he ascended the Potomac aways and recorded on his map what Indians reported of its course beyond his
range. He discovered the Patuxent River and the cliffs of Calvert,
which he called "Richard's Clifts." On the way north he missed
West, Rhode, South, Severn, and Magothy rivers, but found the Patapsco. Evidently he penetrated what is the Inner Harbor of modern
Baltimore and noted the hill later called Federal Hill: "For the red clay
resembling bole Armoniac, we called it [the river] Bolus."[5] At the
head of the Chesapeake Bay he was disappointed that he failed there
to reach the Indies. The year before he had read instructions from the
London Company about choosing a navigable waterway running well
into the country, preferably in a northwesterly direction, "for that
way you shall soonest find the other sea [that is, the Pacific Ocean]."[6]

Three years later (1611) the London Company and its settlers received a gift richer than the gold of the Indies that Smith sought. It came with the immigrant John Rolfe, was imported from the West Indies, and it was, of all things, a simple plant—tobacco. This plant flourished in the leaf mold-rich soil of Virginia. Within a decade everyone there was planting and harvesting it.

Back in 1586 Sir Walter Raleigh had received a tobacco pipe by way of Spain. Shortly afterwards doctors latched onto tobacco as herba panacea, a cure-all. Other people called it herba santa or "devine tobacco." The upper classes of the Old World took it up. And Virginia could—and did—feed the habit. With the news of how much money could be made growing tobacco there, the colony attracted hordes of English people, most of them unfit to live on a frontier. Still they came, and by 1640 all the best land on the James River up to the falls (the site of Richmond) had been taken up. Evidence of the appetite for arable land appears in an advertisement in the *Virginia Gazette*, Williamsburg, August 29, 1771:

A plantation for sale in King and Queen County, just below West Point, known by the name of Gough's Point, containing two hundred acres. . . . on which is a good Deal of Marsh that with little Expense might be made a valuable Meadow.

Virginians, however, soon had competition. Although their company had received three successive charters from the English Crown, their boundaries in Virginia proved amorphous to the king. The terms of the three documents pushed the boundaries out to include parts of North Carolina, all of Maryland, and parts of Pennsylvania. Then on June 20, 1632, King Charles I granted to Lord Baltimore the tract of land that Baltimore called Maryland. Who could question His Majesty's absolute right to give away the same Bay twice?

Here then arose a second colony, and one whose settlers, first arriving on its shore in 1634, sought gold too. They came to an inlet from the Potomac River, St. Mary's River, and called the settlement St. Mary's City. That name seems ironical to us because of its expectation of urban distinction in the tidewater where no real city spread until 175 years later. St. Mary's City turned out to have a better site than Jamestown's because it stood above its river and was well supplied with fresh water.

Colonial rules, set by proprietors and kings, encouraged the founding of port towns. And the authorities chartered many all around the Bay. They and their merchants clearly hoped for new versions of London and Bristol. In 1705 an advertisement appeared in London called "A Plain and Friendly Perswasive to the inhabitants of Virginia and Maryland for promoting towns and cohabitation": "Cohabitation would not only employ thousands of people . . . others would be employed in hunting, fishing, and fowling, and the more diligently if assured of a public market."[7]

The planners concentrated on towns as conduits for the money that was to flow into public and private coffers. What they got instead were headaches. The first Bay settlement, Jamestown, according to ample evidence, seemed always about to collapse.

Government officials persisted in authorizing the establishment of towns. Most efforts failed. When an earlier Baltimore, laid out on the banks of the Bush River in Harford County, Maryland, did not develop, it was replaced by another Baltimore in 1729 on the present site. Again the Maryland colonial assembly took great pains in 1744 to establish Charlestown (in Cecil County on the Northeast) by giving special privileges to owners of waterfront lots.

The assembly offered use of a town wharf and warehouse, free use of public squares, spring and fall fairs, and permission to lease marsh areas in town to anyone to drain. Similar lures succeeded better at Georgetown, Maryland. The commissioners for the new District of Columbia published regulations in July 20, 1795 permitting proprietors of water lots to build wharves "as far out as they think proper" (but not injuring the channel).[8]

Although public policy encouraged urban development, no town had developed by 1662 in the Virginia colony, a place that then had about forty thousand population.[9] The reason for this failure lay in the economy of tobacco growing on far-flung fields of plantation owners that shipped from riverside docks nearby. No planter required a town. But English colonial officials wanted Maryland as well as Virginia to be like New England, a town-based culture. London never faced up to the geographic and agricultural realities of the Chesapeake region. A contemporary comment illuminates the problem: "No country in the world can be more curiously watered. But this conveniency that in future times may make her like the Netherlands, the richest place in all America, at the present I look on the greatest obstacle to trade and commerce."[10]

Chesapeake residents did not need cities then. What they did need was land, lots of tobacco-growing land. And they needed waterways to carry the pungent-smelling cargoes abroad. They found both land and water aplenty. Father White, accompanying the first Marylanders, called the Potomac River "the greatest and the sweetest I have ever seen. The Thames is but a little finger to it."[11] There in southern Maryland the settlers made a good beginning. They avoided the strange fluxes and agues that had afflicted the Virginians earlier. And they enjoyed the abundant game and fish. They learned from the Indians how to grow maize and sweet potatoes. Because the act of religious toleration admitted Quakers, some of that industrious sect had settled around the Bay by 1650, particularly in what is now Anne Arundel County. In 1649 a group of Puritan exiles from Virginia founded Annapolis (at first called Providence), and that settlement became a port of entry. Thirty-five years later it was made the seat of royal government, the proprietorship of the Calverts having ended.

Annapolis did not become much of a town, although as a colonial capital in the eighteenth century it had its pretensions like Williamsburg, Virginia's second capital (1699). We have but to look at population figures. Writing in 1782, Thomas Jefferson reported that Williamsburg never exceeded a population of eighteen hundred. Annapolis had about the same. Richmond, Virginia's third capital and unincorporated until 1805, had only eight thousand in 1820, but by that date Baltimore had zoomed ahead of all tidewater towns (for reasons that we will explore) to sixty-two thousand.

Baltimore stood out as a port as early as 1785, when a visitor noted, "Alexandria [Virginia] has made considerable advances since 1778, but afforded no comparison in its progress, to its vigorous rival, Baltimore."[12] Even moving the new nation's capital to the shores of the Potomac (1800) failed to create much of a city there for years. Note that the largest marsh in the District of Columbia was reclaimed only after 1900. It occupied part of the Mall.

The reason for the small size of the towns rested on tobacco. Back in 1640 half of the English settlers lived in the Massachusetts Bay Colony. Of the other 15,000, about 8,000 lived in Virginia, and 1,500 in Maryland. Twenty years later 26,000 whites and 950 blacks lived in Virginia, and 7,600 whites and 750 slaves lived in Maryland. During the next fifty years the number of slaves increased astronomically, all hands needed to grow tobacco.

In 1724, an Englishman, the Reverend Hugh Jones, published in his *Present State of Virginia* an observation that applied to Maryland as well:

Tobacco Road, Richmond, 1845. A WPA-sponsored mural in a series about Richmond, Virginia. Such roads, called "rolling roads," were found in many communities around the Bay—we still have many Rolling Roads. Courtesy National Archives.

No country is better watered, for the conveniency of which most houses are built near some landing-place; so that anything may be delivered to a gentleman there from London, Bristol, etc. with less trouble and cost, than to one living five miles in the country in England. . . . Thus neither the interest nor inclinations of the Virginians induce them to cohabit in towns; so that they are not forward in contributing their assistance towards the making of particular places, every plantation affording the owner the provision of a little market.[13]

The little market resembled a company town, with the head of the company, the planter, then fulfilling many roles—farmer, judge, merchant, doctor, host, and—always—aristocratic lord of the manor in the English mold. We do well to look at him closely. He was at the center of economic, social, and political events in the tidewater. In him we find influences on the governing of uses of the Bay. He served as legislator, he acted as sheriff, he decided as judge.[14]

Economically speaking, the planter and his tobacco crops dominated the Bay region. These Virginia and Maryland planters ruled over kingdoms. George Washington owned a mere 8,000 acres, small compared with the 63,093 acres of Robert Carter, of Nomini Hall, Virginia.[15]

With five thousand acres and the work of slaves, the planter could afford to live like a lord. (Even fifty acres of tobacco land supported a family in the eighteenth century.)[16] With money to live like a lord, the planter did ape the social life of the English gentleman of status and large estate. And like that gentleman, he could tail his estate to insure keeping the land in the hands of a few. On his land he built and maintained a great house. With plenty of credit in London, he bought luxuries to adorn it. He also bought culture. Though the level must have been superficial at first, the culture of the tidewater deepened by 1750 to shine in what has been called the golden age of the Bay aristocracy.[17]

That golden age shaped major founding fathers in the creating of the United States, including Washington, Jefferson, Madison, and Monroe. Any Bay history with a humanist approach then must record the leadership of planters like these men. George Washington exemplifies the early Virginians' way of settling regional conflicts. Note that he too was an English gentleman, willing to accept change but possessed of traditions that became the foundation of the United States. He was loyal to his parish and to his Chesapeake fiefdom. No wonder he presided over a constitutional convention that codified the federal spirit and also expressed support of states rights.

It is significant for our present history to note how Washington sat down with a handful of other Chesapeake men in Alexandria and at Mount Vernon to solve major conflicts about Chesapeake waters that had long disturbed Marylanders and Virginians. He was not yet president when he and the group worked out the historic Compact of 1785. Their decisions freed Marylanders from the threat of being forced to pay tolls to Virginia at the entrance to the Bay. They also gave Virginians the long sought right to fish the Potomac River, owned entirely by Maryland. In another historic decision but of less lasting duration—one hundred years—this compact moved the boundary between two states. The line set in Lord Baltimore's charter of 1632 had begun well south of the mouth of the Potomac and below South Point of Smith Island. The new line gave much more of the Bay to Virginia and included what is reported as twenty thousand acres of oyster bars.[18]

Chesapeake traditions behind this compact reached through it to the whole nation. Because the meeting of regional leaders at Alexandria and Mount Vernon went so well, the committee members decided that other quarrels between states could be settled by bringing together delegates from all thirteen states. The first call sent out by Virginia failed to bring a big enough delegation to Annapolis. But the next call was to meet at Philadelphia in the summer of 1787. Out of that meeting came the United States Constitution and the inauguration of Washington as first president of the new United States. More than a few historians trace modern American democracy back to Chesapeake Bay and not to New England.[19]

Even when commerce went to town, leadership stayed in the country. But in the nineteenth century towns did grow, and the story of the growth of Baltimore, the largest of them, will illustrate how and why.

Baltimore first made its reputation as a brash town on the Chesapeake Bay. Other tobacco ports, of course, existed before the chartering of Baltimore in 1729. This new town shipped little tobacco as it spread across land owned by the great Carroll family and in time surrounded the plantation house of Charles Carroll the Barrister, Mount Clare. That mansion, which still stands, received guests like Washington and Lafayette in the eighteenth century.

Washington boasted that "no estate in United America [was] more pleasantly situated" than his Mount Vernon. In this painting by Benjamin Henry Latrobe (the only known picture of Washington drawn from life in this setting) Washington is shown scanning Potomac shipping with a spy glass. Seen here in July 1796, the Washingtons are entertaining Lafayette's son, seated next to Mrs. Washington and the tea urn. Also present are Nellie Custis, a grand-daughter, who stands posed like Flaxman's Helen of Troy from a contempor-ary edition of the *Iliad*, and Master Lear, the son of Tobias Lear, Washington's secretary in charge of affairs at Mount Vernon. Latrobe noted that "Down the steep slope, trees and shrubs [are] thickly planted but kept so low as not to interrupt the view but merely to furnish an agreeable border to the extensive prospect beyond." Courtesy James W. Tucker.

Those early guests stopped at Baltimore as others less lordly did because the town Baltimore grew up at a crossroad. It happened to have a harbor on the main overland route south from Philadelphia. More significantly, like Richmond, Virginia, it stands on the fall line of its river. The significance lay in the rush of water in streams, or falls as they were called, that powered the mills. Probably the first mill was erected by Jonathan Hanson about 1710, although in 1661 David Jones bought 380 acres of the stream named for him, Jones Falls, and may have had a mill there. A century later, in 1805 the area within a twenty mile radius of town held sixty mills. From them poured so much flour that that region held the world record. Clearly the economy of this piedmont region then rested on wheat and not tobacco.[20]

With creation of the Baltimore and Ohio Railroad in Baltimore (1828), grain came from the Middle West to be distributed around the world. But grain and flour are only part of the lucky combination that made Baltimore second only to New York in size and importance by the time of the Civil War. Another reason for growth was the metal industry. Back in the eighteenth century, enterprising men like the Carrolls used local deposits to develop one of the largest iron productions in the British Empire. Entrepreneurs also built industries from local copper and chrome deposits.

The third and fourth reasons for the rise of Baltimore lie in the waters, shipping and shipbuilding. An English Quaker family named Fell, one of them a shipbuilder, bought land on deep water (now Fells Point), and inaugurated an industry and a port. Named for an Irish port, Baltimore received immigrants happily. One of them from the north of Ireland, Dr. John Stevenson, often receives credit for the idea of shipping the good Maryland flour abroad. Trade burgeoned. By 1800 Baltimore had become the gateway to the West, the South, the Caribbean, and Europe.[21]

Still, Baltimore was a city-come-lately among East Coast ports. It had only five thousand population in 1770 when Philadelphia ranked as the second largest town in the British Empire. No wonder Congress complained about Baltimore when, in 1776, it took refuge on Baltimore Street from the capture of Philadelphia by British troops. The little town's shipyards, nevertheless, produced sturdy warships for Congress.

After the Revolution the fast schooners of Baltimore carried its name around the world—the Baltimore clipper—and brought wealth and crowds of immigrants. The town's privateers created the "Nest of Pirates" that the British feared and sought to destroy in the battle of Baltimore at Fort McHenry, a climax of the War of 1812. The citizenry rose as one to rebuff the attack in September 1814. They also sang about the victory in the song written by one of them, "The Star-Spangled Banner," and they named streets after its heroes. To this day, Baltimore remains the only major American port never occupied by a foreign power.

Clearly the citizens of the town had guts as well as enterprise. They created first the clipper ship and then the Baltimore and Ohio Railroad. Both modes of transportation helped double the population every decade. Chief molders of the city's character appear to have been businessmen, of the Venetian stamp,[22] as one memorialist called them. Shrewd and aggressive, they ruled the world trade routes as well as the city. Yet part of the city's character yielded to the civilizing waves of gentility and genius—and of women—that washed over the town between 1800 and 1860. It makes a town special to have held within its bounds the first American saint, St. Elizabeth Ann Seton, as well as artists of the calibre of Benjamin Henry Latrobe, the architect, and the Peale family of artists and scientists, and Edgar Allan Poe.

A less attractive side of the city's character caused Baltimore to be called Mobtown. At Commerce and Pratt streets was spilled the first blood in the Civil War, when a mob attacked troops from Massachusetts who were rushing to defend Washington. Baltimore remained divided between North and South, a price paid for its location. Until the end of the war, it felt what the state's anthem called the despot's heel against its windpipe.

After the Civil War, trade in the port of Baltimore still benefited from the harbor's being 150 miles nearer the West Indies and South America than was New York. Among the imports were bananas, sugar, spices, and coffee. Baltimore had its own "Coffee Fleet." Guano came in too to be used in fertilizer, a product that also used local oyster shells in its manufacture. These shells came from oyster canneries that lined the northern waterfront. Canning oysters developed in Baltimore partly because of abundant supplies nearby, partly because of cheap labor, both European immigrant and freed black. Another reason lay in a number of inventors, including Thomas Kensett. These men invented the machines for making and sealing the tin cans.

View of Baltimore from Federal Hill (1850) by E. Whitfield. At this time
Baltimore ranked as one of the largest cities in the nation, and one of the
busiest as the number of smokestacks and ships shows. At the far right a
forest of masts marks Fells Point, shipbuilding center and the original deep-
water port of Baltimore. Courtesy the Library of Congress.

Before 1900 Baltimore became the nation's largest canning center. Processing other products of the Bay region—vegetables and fruits shipped to Baltimore docks—pushed the industry so much that by 1890 canning employees numbered twenty thousand. Left from that era are companies like Crown Cork and Seal and Continental Can that grew up as auxiliaries to the canning factories.[23]

Let's pause here at the economic successors of canning Bay products to send around the world, and look back over three hundred years. In 1607 Chesapeake Bay held out false promise as a route to the East. Settlers followed the disappointed explorers to create a tobacco aristocracy dispersed along its shores. Later came the port cities, preeminently Baltimore, to serve as points of transshipment. All the while, settlers and their successors exploited the waters of the Bay regions. For them the quality of water was not a problem, but a solution. Besides shipping, water was used for drinking, as a sink to dispose of wastes, and as a fishing ground. The problem arose when these uses came into conflict.

# FOUR

# A Well, a Sink, a Pesthole

The Back Basin [of Baltimore in the 1890s] received the effluence of such sewers as existed and emitted a stench as cadaverous and unearthly as that of the canals of Venice.... It continued to afflict the town until the new sewerage system was completed, and the Back Basin was reduced to the humble status of a receptacle for rainwater.
— H. L. Mencken (1941)

The settlers on the shores of the Chesapeake first demanded pure water to drink. Not all the imported beer and sack could substitute. Planters and townsmen alike had trouble finding and retaining sources of good water. As with so much else, we can begin with Captain John Smith. On the Eastern Shore at one point in his explorations of the Bay he and his crew found only a puddle from which to fill "but three barricoes.... We digged and searched in many places, but before two daies were expired, we would have refused two barricoes of gold for one of that puddle water of Wighcocomoco."[1]

## OBTAINING A WATER SUPPLY

When springs were lacking, the inhabitants dug wells. They also dug latrines and, later, cesspools. Luckily, the hilliness of Richmond and Baltimore permitted good draining of storm water and of domestic and industrial wastes up to a point. Also, for years Baltimore's lusty streams rushing through the hills provided both drinking water and a channel for wastes to enter tidewater. This mixture of uses inevitably led to trouble.

It led, moreover, to attempts not only to provide a supply but also to protect the drinking water in towns. In 1792, for example, the Insurance Fire Co. of Baltimore was authorized to sell shares to obtain capital for supplying the town with a reservoir and with pipes to conduct the water.[2] Again in 1800 the legislature gave that town permission to construct a water supply system. We must keep in mind that Baltimore's population was just about doubling every decade after the American Revolution. By 1809 the town's natural springs were being encroached upon by the growth, "threatened with destruction by other improvements."[3]

Central Springs on North Calvert Street (the present site of Mercy Hospital) was incorporated, as soon were two others, Clopper's in the south of town and Sterret's in the east. Like the water company in New York City (from which evolved Chase Manhattan Bank), Baltimore's water company became more intent on other money-making ventures than on providing streams of clean water. The January 15, 1830 *Baltimore Gazette and Daily Advertiser* carried a report of the town's water committee asserting that that company provided only for the central section of town. Besides, ran the complaint, more than half the year the color and consistency of the water made it unfit for any use. And sometimes there was no water at all. This news report went on to say that the water company was "chaining down our fellow citizens to a prolonged subsistence on this muddy substitute for the pure element." People lived in fear of resorting to "pumps which we have found it necessary to abandon, and imbibe disease and putrid water together." Almost as bad as disease was the threat of having to "relinquish the design of multiplying our houses, our streets, and our people." For a booming town that threat must have hit hard.

Norfolk fared even less well. From its earliest days visitors noted that the water was brackish and, some said, unpalatable. A public spring was located near Main and Church streets, but the river itself provided "the place appointed for the public laving" (appointed by the Norfolk Council). So bad was the taste of the drinking water from pollution and brackishness that in the summer of 1800 a man who owned a good well on Briggs' Point, Johnny Rourke, peddled water from his "tea-wagon." Fifty years later most of the drinking water came from cisterns. And only after the Civil War did the city begin to tap Lake Drummond and Deep Creek for good, pure water.[4]

Authorities took steps to prevent contamination of water supplies. In fact, an early use of the word "pollution" appeared in a provision of an act of the Maryland legislature establishing the Baltimore Water Co. in 1808.[5] A fine was imposed on anyone polluting the Jones Falls, the source of drinking water, between the pumping house and the mill. This stream evoked plenty of controversy as a source partly because its supply ranged from trickle to flood. The Baltimore Water Commission in 1852 rejected a report urging the use of the Gunpowder River as its source. Instead, still drawing on the Jones Falls, they built a system complete with classical temples disguising wasteweirs and gatehouses. But it was in vain, for the supply proved inadequate. Worse still, the absence of filtration meant contamination and typhoid epidemics. Soon came new waterworks drawing on Gunpowder River. Ultimately, Baltimoreans were compelled to send fifty miles to the Susquehanna River for water.

As late as 1886 the Maryland legislature had to reiterate that there was pollution of the water supplies.[6] In 1874 it was made illegal to throw dead animals and other substances into the Potomac River.[7] That tributary provided drinking water for many settlements, including the District of Columbia. The report of the Baltimore Water Commission in 1853 argued against building an open canal to the reservoir from the source, because "water is exposed to all the filth and dirt which anyone chooses to throw into it."[8]

A classic case of conflict about the use of water for drinking and for the disposal of wastes was decided in 1882 when the mayor and city council of Baltimore sued the Warren Manufacturing Co.[9] The defendant's cotton factory, it was complained, dumped "divers injurious ingredients and substances" into the Gunpowder River that had been dammed for reservoirs by Baltimore city. The resulting pollution contaminated the drinking water of Baltimore. The court held that the privies and hogpens near the factory did indeed pollute and had to be removed. But the city was denied relief against the manufacturer since it was unable to prove the nature of the pollutants emanating from the manufacturing processes of the company.

A sampling of the case law in Virginia during the nineteenth century reveals that the courts had to deal mostly with mills and milldams. Courts decided not only about competing rights, but they also addressed pollution issues.

For example, in 1828 a milldam across Little Creek in Nottoway County, Virginia, caused stagnant waters. The commonwealth sued, alleging that the dam was a public nuisance. The plaintiff lost, having failed to show how stagnation affected a public highway or some other place in which the public had such special interest.[10]

Again, in 1833, a court dealt with a case of stagnation caused by the construction of a mill and milldam, this one on Deep Creek,

During the nineteenth century, public springs and corner pumps supplied
water in Chesapeake towns. This pump stood at the corner of Gorsuch
Avenue and Loch Raven Road, Baltimore, as late as 1936. Courtesy Enoch
Pratt Free Library.

Virginia. The Maud family suffered "severe bilious diseases" every year, and several "intelligent physicians" and others from the territory testified that the diseases were "justly imputed to the effluvia from the stagnant waters of the mill pond." At the washing away of the mill, a suit was brought to prevent the owner from rebuilding. The latter argued that he had built under statutory authority and had a statutory right to rebuild. The court of equity asserted it had the authority to "prevent the destruction of health and life," and it distinguished between danger to health and danger to fish and navigation. The court ruled that it will allow experiments with the latter, but as to the former "The law tolerates no such experiments at the risk of human life."[11]

Finally, at close of the nineteenth century, the Virginia Supreme Court of Appeals dealt with the problem of pollution itself. In *Trevett v. Prison Association of Virginia*[12] the court ordered that damages be paid to the plaintiff who had used the waters of a stream for domestic purposes and for a number of cows whose milk and butter commanded the highest price on the market. The defendant operated a school for young criminals which emptied "refuse water, urine, and excrement" into the stream, thereby destroying the plaintiff's dairy business.[13]

Health certainly posed the major issue connected with water quality in the first 275 years of Bay history. The chief enemies to good health, contagious diseases, were reported in newspapers every August. Epidemics as well as individual incidents fill letters and journals with appropriate alarm. For example, in the fall of 1762 Charles Carroll the Barrister wrote from Annapolis, "I this summer made an Excursion as far as Boston in order to Escape my Troublesome annual visitant the fever and Ague [possibly a form of malaria] but had not been returned to Annapolis four days before I was seized with it in a more violent manner than at any of its former attacks and it still keeps possession of me."[14] For another instance, Colonel Landon Carter recorded in his diary for October 3, 1765, "This is a strange ague and fever season. The whole neighborhood are almost every day sending to me."[15]

These neighbors no doubt were sending for medicines and advice. And the advice probably concerned the air rather than the water, because contagion, it was thought for centuries, was airborne. The truth came at the very end of the nineteenth century. Appropriately for our history, two Chesapeake doctors helped fix blame on the water. Although the bacterial cause of cholera wasn't proven until 1884 in Egypt, a generation earlier a Baltimore doctor came close to the truth.[16] Dr. Thomas Buckler attempted a scientific study of a cholera outbreak in the almshouse at Calverton outside Baltimore as well as in that city. Then in 1851 he published *A History of Epidemic Cholera.* In it he cited a case of filth in Baltimore in connection with nine deaths from cholera.

The other Baltimore doctor, Dr. Jesse Lazear, became a martyr to discovering the link between mosquitoes and both malaria (swamp fever) and yellow fever. A fellow worker on the United States Army Yellow Fever Commission, Dr. James Carroll, wrote that Lazear contributed "the two first authentic cases of experimental yellow fever on record."

> Then with a full knowledge of the power of the insect to convey the disease, he afterwards calmly permitted a stray mosquito that had alighted upon his hand in a yellow fever ward, to take its fill, and inject into his system the virus that twelve days later robbed him of his life.[17]

Today an inscription on a plaque in Johns Hopkins Hospital (where he trained) reads,

> With more than the courage and devotion of the soldier he risked and lost his life to show how a fearful pestilence is communicated and how its ravages may be prevented.

The summer epidemic of 1855 in Norfolk well illustrates the horrors of yellow fever. A steamer bound from St. Thomas to New York put into Hampton Roads for repairs. On board were cases of yellow fever, a fact kept hidden until a shipyard laborer contracted the disease. By August the city of Norfolk as well as Portsmouth suffered decimation of population. Reports said that the city was wrapped in gloom. Most people were either confined with sickness or waited on the sick. The number of deaths rose to one hundred daily, and the supply of coffins gave out. The calamity shocked residents particularly since they had begun to believe that their attention to draining streets and using cisterns for drinking water had saved the city from the recurrence of epidemics such as those of 1821 and 1836.[18]

Twenty-three years after the 1855 epidemic, Congress passed the National Quarantine Act. This and other such acts gave to federal agencies some duties that for almost three hundred years had been

Four scientists, all born in the Chesapeake region in the nineteenth century, who have made lasting contributions in the field of public health. *Left, top,* Dr. Thomas Buckler in 1870, an early investigator of the cause of cholera. (His image is superimposed on what at that time was considered a suitable background for a doctor.) Courtesy Maryland Historical Society. *Left, bottom,* Dr. Jesse Lazear, a native of Baltimore who trained at Johns Hopkins Medical School, served on Dr. Walter Reed's Yellow Fever Commission, let an infected mosquito bite him, caught the fever, and died. His death, September 25, 1900, helped prove the cause of the disease. Photo taken in the 1890s. Courtesy Alan M. Chesney Archives, The Johns Hopkins Medical Institutions. *Above,* Dr. William H. Howell in 1900, a founder of the Johns Hopkins School of Public Health, the first school of its kind in the world. Courtesy Alan M. Chesney Archives, The Johns Hopkins Medical Institutions. *Below,* Professor Abel Wolman of Johns Hopkins University, author and internationally recognized authority on water quality, who is still working at age ninety. Courtesy Ferdinand Hamburger Archives, Johns Hopkins University.

performed by local governments. For example, as early as 1801 the Maryland General Assembly had authorized the construction in Baltimore of the lazaretto (a quarantine station) at the northeast edge of Baltimore harbor.[19] This city already had its health police, agents for the commission of citizens, that was set up when Baltimore was incorporated by the state legislature in 1797.

About the time of the National Quarantine Act, states were establishing official boards of health. Maryland's was created in 1874 to guard the sanitary interests of the people of the state and to investigate the causes of disease and mortality and the influence of locality, employment, habits, etc., on health.[20] It was also created to investigate nuisances. Local boards of health were authorized by Maryland's legislators in 1886.

Just four years earlier, in 1882, Dr. W. C. Van Bibber read a paper before the Medical and Chirurgical Faculty of Maryland, later published as "The Drinking Waters in Maryland Considered With Reference to the Health of the Inhabitants."[21] He reported that on January 19, 1881 the hydrant water of Baltimore had tasted and smelled "disagreeable," and that the public was aroused. "The public-spirited proprietors of the *Sun* newspapers" hired a Professor Tonry, and the water board got Professor Ira Remsen to investigate the cause. Tonry gave his opinion that the source, Jones Falls, was polluted "due to the decomposition of the sulphates, held in solution, passing into the sulphites, and setting free sulphuretted hydrogen gas." The bad taste and smell lasted until the middle of April. Citizens using pump water instead of hydrant water were warned by Dr. Van Bibber to be suspicious of their supply also. "When it is known that there are over 80,000 nuisance sinks and wells at various levels within the city limits of 9,600 acres, no one will wonder that the pump waters within this area, and even beyond it, should be contaminated."[22]

Just how far attitudes had changed we can see by comparing these steps with the comment of a civil engineer, Charles Varle, writing in 1833, only fifty years earlier. After praising what he called the efficiency of the health police of Baltimore, he noted the calamitous ravages of cholera in 1832. A total of 3,572 people died out of a population of 80,990. Varle added, "May the great Disposer of events hereafter keep us free from pestilential visitations."[23]

Less pious citizens prodded early officials to prevent the spreading of disease associated with water and waterfronts. The fever, it appears, was more terrifying than fire or flood. One gadfly, Dr. Thomas

The original lazaretto (now demolished) built by Baltimore in 1801 as pest hospital and quarantine station. In the foreground are navigational buoys brought in for refurbishing in 1940 when the building was used by the Coast Guard. Courtesy the *Baltimore Evening Sun.*

Buckler, published a map in 1851 (see page 43) showing the areas of
remittent and intermittent fevers around the Patapsco River. These
were extensive. Buckler had his theories of the causes, just as other
people did: he blamed uncleanliness generally and decaying refuse
and garbage in particular. Darkly he prophesied that "the germs of an
epidemic introduced into the polluted atmosphere surrounding this
festering pond [the Inner Harbor of Baltimore]" would spread. His
solution combined ingenuity with a lucrative angle. It was simple:
shave off the hill adjacent to the Basin, Federal Hill, and fill in the
Basin for building lots.[24] (Had city fathers agreed with him, Baltimoreans today would have no Inner Harbor, the centerpiece of the city's
renaissance.)

Thus Buckler would have rid the city forever of what he called
"the most unwholesome, disease-engendering and pestilential condition of the Basin and Docks." He also would have eliminated "the
stagnant mill-ponds along the Jones Falls, to say nothing of the
putrid, seething and malarial condition of the tidal portion of the
Falls." In addition, he suggested deepening the falls, so that "the tide
may ebb and flow as high up as Madison Street bridge" and in its
scouring remove the causes of disease.[25]

Back in 1797 another physician, Dr. John Davidge, a leader in
Maryland medical circles, declared that yellow fever was not contagious. In Philadelphia the well-known Dr. Benjamin Rush agreed, but
few other people did. People noticed that this disease for a period
always started at Fells Point, and usually near Smith's Dock. Some
noted that just east of the Point and between it and the lazaretto
(quarantine station) stood stagnant ponds. That was true during the
epidemic of 1819.[26]

After 1822, statistics show a decline in deaths in Baltimore as
yellow fever and malaria diminished. The reason lay in better drainage and in less commerce with the West Indies. Confirmation comes
from an authority on water quality, Abel Wolman of Johns Hopkins
University:

> The effects of epidemics in the early days of the Republic on
> pushing forward installation of public water supply and sewer-
> age facilities can hardly be overestimated. Yellow fever and
> cholera between them created havoc decade after decade with
> important results so far as municipal housekeeping was con-
> cerned. . . . The increasing public demand for sewering and
> draining cities to avoid nuisance and to gain convenience di-

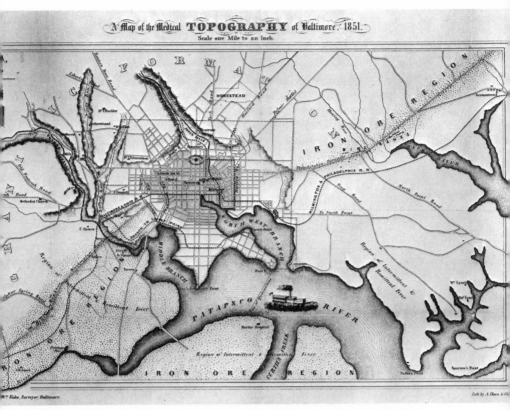

In 1851 Dr. Thomas Buckler published this map to show locations of the frequent outbreaks of fevers. Note the reservoir, next to Jones Falls and north of the Washington Monument, which was the source of drinking water. The lazaretto can be seen on the map just above the steamboat. Dr. Buckler's house, marked in the upper left, was just north of the almshouse where he did his research. Courtesy the Library of Congress.

minished remarkably the incidence of yellow fever, due primarily to the elimination of the breeding places for mosquitoes. In similar fashion the epidemics of cholera in 1832, 1843, 1849, 1854, 1865 and 1873 brought about a certain amount of recognition of the probable relationship between this disease and the inadequate and unsafe methods of the disposal of household wastes, and undoubtedly stimulated the establishment of drainage and sewerage systems in the larger communities.[27]

An urban geographer recently wrote that in this period a learning process was at work resulting in what she calls the creation of "new institutions of collective response." The series of epidemics, she hypothesizes, called attention to certain "foul and filthy spots and spurred the increase of regulatory power."[28]

People believed that the shorelines of harbors threatened disease. Public opinion, therefore, forced attention of town and county officials to get rid of marshes and low ground around settlements. In 1766 an "Act to Remove a Nuisance in Baltimore town" denounced in vivid language "a large miry marsh giving off noxious vapors and putrid effluvia."[29] Landowners were given four years to wall off this blemish on the waterfront and fill it in.

So sensitive indeed were the noses of the nineteenth century that writers vied with each other to describe the various perfumes wafting up from the Basin in Baltimore. One called it "hellbroth."[30] It was "a bubbling cesspool" remembered Abel Wolman, a man whose memory reaches back to 1900. He recalls returning at night about 1910 on the boat from Tolchester and knowing when he was within five miles of the city by the odor.[31]

Early nineteenth century medicine sought the cause of diseases in the fumes given off from putrescent vegetation as well as from marshes and lowlands. A contemporary historian concludes that Dr. Benjamin Rush of Philadelphia "tried to improve the public health in Philadelphia by such common-sense expedients as sewage disposal, pure water, and clean streets."[32] In other cities similar analyses appeared. Among reports of the Baltimore Public Health Department for 1825 appears a note by a prominent doctor, Thomas E. Bond, who was called "Consulting Physician": the situation on Washington Street, Fells Point, "threatens to produce unfriendly effects on the health of neighboring inhabitants [because] the fall is insufficient to carry off the water with sufficient celerity. . . ." He added that below the apron [of the planks of the south end of the street] a "very dangerous nuisance" arose because "water ... becomes stagnant, and emits at this season a very disagreeable smell and will in the hotter season of the year be a fertile source of deleterious exhaltation."[33]

## DISPOSING OF WASTE

The brackish waters of Chesapeake Bay, while unsuitable for drinking, were an ideal sink for waste disposal. Inhabitants flushed their sewage and refuse into the receiving waters of the estuary. In Norfolk, wrote traveler Mederic Moreau de St. Mery, in 1794, "The sewage ditches are open, and one crosses them on little narrow bridges made of short lengths of plank nailed to cross-pieces."[34] These ditches, of course, emptied into the convenient Bay.

Sixty years later Washington, D. C., appeared "as sour as a medieval plague spot," according to a later historian of the Civil War. She called the sanitary conditions of the crowded capital appalling. "With its river flats, its defective sewage system and its many privies, it had always been odorous in warm weather," but now in wartime, crowded District of Columbia epidemics of disease threatened. "The comatose Board of Health aroused itself to a protest, the Surgeon General made inquiries and the War Department conducted an investigation," she cynically concludes.[35] Not until 1889 were sewers constructed by the Board of Engineers.

In Baltimore at that time, the harbor was ranked "among the great stenches of the world."[36] That city stood out as the last of the great American cities to build a modern system for sewage. It came between 1906 and 1915. Writing in 1912, Clayton Coleman Hall boasted that completion of the system would make Baltimore "the best sewered and drained city in the world."[37] His enthusiasm carried him into error when he added that "the sewage [after processing] when discharged into Back River will be purer than the water into which it is discharged and will be absolutely innocuous to the oysters and fish inhabiting the waters of the bay." Such proved not to be the case.

Baltimore in fact makes a useful example because it became a metropolis of the Chesapeake region. By 1850 the city had grown to be the third largest in the United States with 170,000 population. The 70,000 represented the growth during the preceding decade. In the article about Baltimore in the contemporary *Gleason's Pictorial Drawing-Room Companion* (1855), the writer noted that "The city is built on quite uneven ground, which gives an advantage in relation to cleanliness."

Fifty years later fifteen thousand houses had private lines to the Jones Falls for disposing of raw sewage. But more houses had cesspools that were cleaned regularly by Odorless Excavating Apparatus Co. The population of half a million was too large for what had once been of such great service, the good drainage of the hilly terrain and its streams. The soil became saturated, floods occurred in downtown, and the Inner Harbor smelled to high heaven. The gadfly Dr. Thomas Buckler proposed facetiously that the gases be collected to use in Congress. And he worried, not facetiously, that "other terms of an epidemic introduced into the polluted atmosphere surrounding this festering pond [will spread]."[38]

Dr. Buckler wasn't the only one offering proposals. In the mid-1870s fifteen plans lay on the table of the Joint Standing Committee on the Harbor. In a plan by Milo W. Locke, the water was to be kept in motion and diluted constantly. An editorial writer even seriously proposed building a dam across the mouth of the Inner Harbor. The water brought in at high tide could be held and released at low tide in a flushing operation. The flood might, of course, have wreaked havoc on the ships below.

Another source of pollution came from the draining of the filth from streets and shore. For example, the records of the city of Baltimore, between 1782 and 1797, include a resolution in June 1795 "that Daniel Grant be notified to remove a Nuisance on Public Alley, which is occasioned by his necessary."[39] Other nuisances strewed the alleys—and the main streets. Even as late as 1900 stepping-stones remained in use to aid pedestrians in crossing alleys clean shod above the open drains. (See illustration p. 47.)

Streams like Harford Run and Jones Falls in Baltimore became nothing but canals to carry off the town's wastes. A member of Parliament, Sir George Campbell, commented in 1879, "A peculiarity of Baltimore is that there is no system of underground drainage" and in an understatement worthy of quotation added that this arrangement was "not very agreeable to the senses." But he was not certain that this city's old-fashioned way was not "much more wholesome than [London's] underground system."[40] Why he didn't say. Possibly he was thinking of sewer gas.

A century earlier, Baltimore Town Commissioners in February 1768 had devised a system of draining water from the main street, Baltimore Street. The record specifically states the importance of directing it to the tidewater. The commissioners decided to make a

STREET IN BALTIMORE, MARYLAND, U.S.

As late as 1915 pedestrians crossed open drains on raised stones in street beds.
This engraving in the *London News*, October 4, 1856, shows stepping-stones
across Liberty Street, Baltimore. Courtesy the Library of Congress.

small ditch or canal on both sides of the street, the one on the south
(or harbor side) to lead to Gay Street and thence down Gay Street
through a canal to the waterside. The north side required a more
elaborate route by way of the creek (Jones Falls). Later Ann Street,
running downhill to the harbor, was given a concave shape with a
one-foot rise to create a storm drain, carrying trash as well as sudden
runoffs of water into the Inner Harbor.[41]

Agricultural and industrial wastes also were disposed of in
streams and tidewater. Plenty of running water made this flushing
easy and effective. But even as early as the seventeenth century, town
fathers took care to isolate certain trades from the town center. In
laying out Annapolis, for example, Governor Nicholson imposed an
elaborate baroque plan placing centers of church and of government
on hilltops and creating separate districts, quite distant, for "Brew-
houses, Bakehouses, Dyers, Salt, Soap, and Sugar-Boilers, Chandlers,
Hatmakers, Slaughterhouses, some sort of Fish-mongers, etc."[42]

No account of waste disposal can end without a note on the
widespread coal mining. In mountainous regions bordering tributaries of the Bay, wastes from this vast industry poured into the
network of streams. For example, silt and acid from anthracite coal
were dumped into the Susquehanna River's North Branch in Wyo-
ming Valley, Pennsylvania, and bituminous coal wastes went into
the West Branch.[43]

# FIVE

# Traders and Watermen

The Chesapeake is the Mediterranean Sea of America.
—Publicity (late nineteenth century)

For settlers faced with malnutrition and starvation, the Bay contained godsent foodstuff. Shellfish and finfish were plentiful. This plentifulness of fish was attested to many times in writing as, for example, by George Alsop, a seventeenth century observer:

As for fish, which dwell in the watery tenements of the deep, here in Maryland is a large sufficiency, and plenty of almost all sorts of Fishes, which live and inhabit within her several Rivers and Creeks, far beyond the apprehending or crediting of those that never saw the same, which with very ease is catched, to the great refreshment of the Inhabitants of the Province.[1]

John Smith noted that the male Indians "bestowe their times in fishing, hunting, wars, and such manlike exercises."[2] And Smith's contemporary in Virginia, George Percy, wrote of the very large oysters roasted by Indians, so "delicate in taste."[3] Later the flavor of the seafood also came in for comment, as on October 13, 1765, William Gregory noted in his journal, "I put up at Midton's [Annapolis]. Fine oysters to be had." The next day he wrote, "I set out for Joppa about 12 o'clock and arrived here just 8 o'clock. Pretty large vessels coming up to this place. Good oysters. . . . Gunpowder river runs past here."[4]

## THE BAY AS A FISHERY

The colonial governments in both Maryland and Virginia received from the Crown the right to dispose of the tidewater subject to public rights of fishing.[5] And exercise their right of fishery, the public did. Even in the first century of colonization some waters were overfished. In 1678 the Middlesex court in Virginia acted to conserve the country's fish. Certain residents had overfished, in the words of the complaint, "to the Great Hurte and Grievance of most of the Inhabitants of this Country."[6] A century later, in 1796, the Maryland Assembly passed an act for the preservation of a breed of fish in the Patuxent River: there should be no whipping or beating the water between February 1 and June 1.[7]

Following American Independence the inhabitants of Maryland and Virginia succeeded to all rights previously held by Crown and Parliament. These rights were to be exercised by the General Assembly in Maryland and the legislature in Virginia. The Virginia Constitution restricted this legislative power by prohibiting sale or lease of "natural oyster beds."[8]

Not long after, disputes arose between Maryland and Virginia as to fishing rights. The Potomac River served as the major boundary between the states, and Maryland claimed by virtue of its charter, exclusive rights therein. Virginia was taking advantage of its sovereignty over the capes of the Bay to charge tolls for vessels carrying goods to and from Maryland. In the Compact of 1785 Virginia agreed to give vessels bound for Maryland free passage in return for an agreement by Maryland that the right of fishery in the Potomac was to be equally enjoyed by citizens of both states.[9]

In the Potomac then, fishery was governed by "mutual consent and approbation of both sides." Elsewhere in the tidewater the states went their respective ways. Disposing of the tidewater, subject to public rights of fishery, was the way local governments encouraged commerce. In 1745, for example, the Maryland General Assembly decided that "all improvements, of what kind so ever, either wharfs, houses or other buildings, that have or shall be made out of the water, or where it usually flows shall [as an encouragement to such improvers] be forever deemed the right, title and inheritance of such improver."[10] Then, in 1759, wharves one thousand feet long were built by John Smith, William Buchanan, and William Spear. Without this lengthy wharfage to deep water, Baltimore would not have grown as a port.[11]

From the beginning fishermen put pressure on legislatures to protect the supply of fish by keeping waters open for the passage of fish. Among these regulations, those in *Acts of Assembly Now in Force in the Colony of Virginia for 1769* order that owners of mills, hedges, and stone stops shall make openings for passage of fish. The courts also ruled against obstructing rivers and streams with mill-dams, weirs, and hedges. Since these regulations of course aided navigation as well, much of the wording was general. But many specifically call the obstruction (usually a mill) deleterious to fish.

Examples of the conflict between fishermen and millowners would fill a book. In Virginia, for example, the legislature passed an act in 1748 that authorized clearing rivers where the passage of fish was obstructed,[12] and it went on in 1759 to order owners of mills on the Rapidan to make slopes for the passage of fish because the inhabitants of Culpeper and Orange counties had complained of obstruction. Within the decade, other acts forced millowners on the Rivanna, Hedgeman, and Meherrin rivers also to make slopes.[13] Building mills on the Rockfish River below the forks was prohibited, and one Allan Howard had to tear down his mill in 1761 because it was entirely obstructing the passage of fish.[14] The Maryland legislature also enacted legislation, usually about a specific tributary: the Patuxent in 1802, the Monocacy in 1806, the Susquehanna in 1813, and the Pocomoke in 1862.[15]

In these official actions, governmental officials responded to, and evolved a way of dealing with, conflicts over different uses of tributary waters. Back in the seventeenth century the council and burgesses of Virginia apprised the governor of the "great mischiefs and inconveniences" that killing whales in the Chesapeake "accrew to the inhabitants." The reason: by these killings "great quantities of fish are poisoned and destroyed and the rivers also made noisome and offensive."[16] Fisheries were subject to regulations about polluting also. In 1810 the Maryland Assembly ordered them to remove the pickle used in their operations, to clean their slips, and to remove all fish from shores within ten days of the end of the season.[17]

But fish quantity was the goal. One fishery owner, a Virginian named John F. Mercer boasted in 1797 that he "contrived to land 20,000 [herring] a day." And to his neighbor on the Potomac near Maryland Point, Richard Sprigg, he wrote, "If I had your lease [as well], I could make a fortune."[18] Anne Royall, a Maryland journalist, traveled in the early nineteenth century around the United States and reported on the large fisheries for herring (and shad) at the mouth of the Susquehanna: the fish were taken in large nets "from 108 to 200 fathoms in length spread across the river by boats. The ten fisheries employed fourteen to fifteen men each at fifteen dollars per month, with their provisions, to catch, cure, and pack the herrings in bar-

rels." The author added: "These fishermen make a wretched appear-
ance, they certainly bring up the rear of the human race. They were
scarcely covered with clothes, were mostly drunk, and had the looks
of the veriest sots upon earth."[19]

Like finfishermen, shellfishermen took all they could—an enormous volume in the late nineteenth century—and jealously guarded
their territories. 'Twas ever thus: an anecdote will illustrate. The
enthusiasm for oysters led a group of five seventeenth century Virginians to break the sabbath. When discovered, they had a canoe full of
oysters. But they escaped punishment, and presumably kept the oysters, because the sick wife of one of them had a craving for oysters,
and on Saturday the wind had prevented her husband from going out
and tonging them.[20]

Colonists began making a living by oystering early. And by the
end of the eighteenth century they had organized an industry. They
had borrowed the Indians' log canoe and improved on it for transport-
ing oysters from the rocks, the natural oyster beds. Likewise, they
had appropriated the Indians' rake, joined two of them with a hinge,
and thus invented the oyster tong that continues in use in the twentieth century.

By 1875 these watermen (as they were, and are still, called) were
harvesting millions of bushels of oysters a year. But the supply did not
last: "It is significant to note that while the population of the United
States between 1880 and 1930, increased by 144.8 percent, the per
capita oyster consumption declined by 77.3 percent." In this period
there was an 81.6 percent decline in annual catch. In 1880 Maryland
had 47.5 percent of the Atlantic and Gulf production, and by 1930 it
had only 15.4 percent.[21]

When necessary, watermen looked to the courts to protect their
rights in God's oysters. At the time of the Civil War in *Phipps v.
Maryland*,[22] a defendant was accused of taking oysters from a private
bed. He defended himself on the grounds that there was a common
law right of public fishery. Apparently, however, such rights do not
extend across state lines. A decade after the Civil War in *McCready v.
Virginia*,[23] the United States Supreme Court held that the Virginia
legislature could lawfully exclude a Maryland resident from taking
oysters in Virginia waters.

Watermen also looked to the legislatures. The greatest volume of
legislation in the 1800s concerned fish and oysters, mostly oysters.
State legislatures, particularly Maryland's, tried to control oystering

and to maintain the supply. One student of the subject repeats the legend that ever since the law against dredging in 1820, the Maryland legislators have enacted more laws on oysters and oystering than on all other topics.[24] For example, the Maryland Assembly in 1834 met the concern of oystermen that improper seines in use were covering oyster beds with mud and thus diminishing the harvest.[25] In 1860 the Assembly set seasons for oystering.[26] It also imposed fines for throwing oyster shells in the water. Already—since 1691—laws existed on the books of Virginia prohibiting the "casting of Ballast into Rivers and Creeks" (1691) and requiring the placing of ballast on land and above the high water marker.[27] In 1748 the Virginia legislature authorized overseers to direct the delivery of ballast to the shores.[28] No doubt such laws were passed to aid navigation as well as to protect fisheries.

## NAVIGATION

From Brother Carrera in 1572 on, no one overlooked profit from trade. Many factors contributed to successful commerce in what a nineteenth century puff called "the Mediterranean Sea of America." For one, even if the Chesapeake doesn't exactly match the Mediterranean climate, it nevertheless remains open year round. There is little fog. And as an observer said of Richmond when that city became the capital of Virginia in 1779, it is "more safe and central than any other town on navigable water."[29]

Another advantage derives from the rush of fresh water that spills into tributaries of the Bay from the piedmont falls. In the past that fresh water was used to kill what were called the termites of the sea, the shipworm, or teredo, enemy of wooden ships. The waters of the Jones Falls in Baltimore's Inner Harbor cleansed many a sailing ship.

Profitably and significantly, the sprawling tidal estuary permitted ships to sail far inland. This realization occurred to William Byrd II in laying the foundations of both Richmond and Petersburg in the 1730s. He wrote that "the uppermost landing of the James and Appomatox rivers . . . are naturally intended for marts where the traffic of the outer inhabitants must center. Thus we did not build castles only, but also cities in the air."[30]

Merchants especially worked to promote the navigability of these rivers. They sought the help of both courts and legislatures to keep ships moving smoothly wherever trade beckoned. Various colonial bodies tried to assure removal of floating trees, logs, and limbs; pushing a tree into the water is easier and quicker than sawing it or burning it on land.

From the beginning, Virginia lawmakers focused whatever attention they gave to water on its usefulness as a road. The focus was based on economics, not twentieth century concerns with pollution or siltation. Laws spread on the books that managed improvements to navigation on every major river. When in 1679 a grand assembly met at St. James City, it passed a law authorizing the clearing of rivers near their headwaters of logs and trees. Furthermore, the counties were left to decide when and how to conduct improvements of the rivers to facilitate the passage of ships.[31] Then in 1680 the legislature authorized appointment of county surveyors to clear waterways. Cited was the loss of boats because "logs, trees, etc." were not cleared. A fine of five hundred pounds of tobacco was imposed for obstructing navigation.[32]

Again in 1722 the legislature passed an act for the "More Effectual Clearing of Rivers and Creeks." In the preamble to the act, the legislature declared that "many of the rivers and creeks of this colony are stopped and choked up by the fall of trees, stumps, and rubbish ... and hedges are made across the same whereby the passage of . . . vessels is hindered and obstructed . . . to the great damage of the inhabitants ... hindrance of their trade and commerce." The counties were[33] then authorized to contract to clear the streams. A later act—in 1726—especially noted that the debris thrown into the water not only obstructed navigation but also caused bridges to break down and get washed away.[34]

Since shipping tobacco dominated trade in the first two centuries, certain acts specifically referred to the more convenient transport of tobacco. For example, one hundred pounds, appropriated at the time, was to clear the Fluvanna River of rocks.[35] In Maryland also the legislature and courts helped keep channels open. In *Harrison v. Sterett*, decided in 1774, the provincial court of Maryland granted relief to one waterfront lot owner against a neighbor who was constructing a wharf which would have restricted his water access.[36] And in 1768 the Maryland Assembly passed an "Act to Prevent Obstructions in the Patomack and Monocacy Rivers" caused by fish dams.[37]

Townspeople insisted that leaders manage shipping in the harbors. Ports around the Bay revived the ancient English post of port warden for the purpose. The warden's duties included conducting

By the end of the eighteenth century Hampton Roads, Virginia, was a prosperous trading center with several harbors, among them Norfolk. This sketch and watercolor drawing comes from the journal of Benjamin Henry Latrobe when he came from England as a young, unknown architect in 1796. Courtesy the Library of Congress.

surveys of channels and keeping them clear and clean. He saw to removing obstructions and annoyances. The Maryland Act Appointing Wardens for Baltimore Town in 1783 (just before the incorporation of the city) permitted the wardens to pass regulations and ordinances preventing injury to the harbor from wharf construction, discharge of ballast, or any other cause.[38] They were specifically asked to approve all wharves that divert the channel or obstruct navigation.

In the Compact of 1785, Maryland and Virginia agreed that their respective citizens could share navigation rights over the Potomac River, and that Virginia would give ships bound for Maryland free access to the Bay. An act of the Maryland Assembly in 1791 authorized commissioners of the city of Washington to enact regulations governing the discharging of ballast from ships in the Potomac above the lower line of the District of Columbia and Georgetown and in the eastern branch.[39]

Ballast discharging bothered Baltimoreans too. But, more important, a persistent concern of Baltimore's merchants was the shallowness of the Inner Harbor, where the wharves were both numerous and long. Wharves were numerous because the chief merchants ruled from there, and the wharves were long because they had to reach out to the seagoing ships. The shallowness was increased by the deposit of sediment washed down by the Jones Falls into the narrow basin. An act of the Maryland Assembly in 1791 authorized the deepening.[40] And after the War of 1812, the leading politician and a merchant in town, Samuel Smith, persuaded the federal government that the ships sunk in the channel as defense against the British attack on the city in September 1814 had to be removed at the federal government's expense. The funds obtained not only raised the ships but also deepened the channel.

That improvement came at just the right time. The first steamboat on the Bay, the *Chesapeake*, was built in Baltimore in 1813. Soon the nineteenth century steam revolution filled the waters with trade. Just within the Bay region a number of steamship lines connected major ports as well as small landings like Lodge Landing on the South Yeocomico River, Virginia, with major ports. For one hundred years and more, everyone knew the routes and names of such lines as Weems and of steamships like the *Lancaster* and the *Maggie*—and others with names of females best known to shipowners. Clearly, the Bay world shrank with the first steamboat.

Running the Rapids of New River, Virginia. A wood engraving after a drawing by W. L. Sheppard, in *Harper's Weekly*, February 21, 1874, shows how rivers were used to bring products down to steamboat landings, and then to markets. Courtesy the Library of Congress.

Because of its speed and certain schedule, steamboats replaced sailing packets on the route from Baltimore north to stage connections with Philadelphia. After the Chesapeake and Delaware Canal opened in 1829, boats linked the two cities. The *Chesapeake* of course was a harbinger—and not just for Bay routes. Soon coastwise steamers arrived, and then overseas ships. Their histories fill volumes. By 1972, the date our story ends, oil tankers created floating islands as they waited to enter the Baltimore Channel.

From those days of sail to our own huge tankers and container cargo ships, Bay ports flourished. How great a volume of trade these ports brought can be told from the rankings of Baltimore and Hampton Roads. These two ports hogged the trade after the Civil War. Before the war, however, docks upstream served the hinterland well. Courts and legislatures saw to maintaining channels to upper estuary towns like Richmond and Georgetown.[41] Virginia and Maryland legislators were forever opening and extending navigation beyond the settled districts.

In 1817 Maryland authorized the use of lottery revenues to improve navigation,[42] and, in 1814, created the Potomac Company to extend and improve navigation on the Potomac.[43] Earlier it acted to remove bars, shoals, and other obstructions from the Anacostia River near Bladensburg.[44] But the act did little good because the silting up continued, and Bladensburg remained (and remains) a backwater isolated by silt.

An unfortunate effect of settlement on the Bay's shore was erosion. Early settlers wrote of soil erosion and pollution, and the late seventeenth century visitors noted erosion and the muddiness of freshets.[45] By 1800 silting had destroyed the channels leading to prominent tobacco ports at Joppatowne, Port Tobacco, and Upper Marlboro in Maryland. On the Gunpowder River, Joppatowne once took an eight-foot draft. But just between 1848 and 1897 seven thousand nine hundred cubic yards of sediment were deposited in the upper part of the Gunpowder River estuary.

To blame was careless use.[46] In 1753 the Maryland Assembly passed an "Act to Prevent Injuring the Navigation of Baltimore-town and to the Inspection House at Elk Ridge Landing on the Patapsco." The legislators noted that by opening and digging into banks of that river, large quantities of earth and sand were washed into the water. They also made it unlawful to put earth, sand, etc. into navigable branches unless well secured.[47]

# The Great Eastern
# EXCURSION!

## THIS MAMMOTH STEAMSHIP,
### The Wonder of the Age, will arrive at Norfolk and be open for the reception of Visitors

## ON SATURDAY, AUGUST 4.

In order to enable the Washingtonians and others an opportunity of visiting this noble vessel,

## THE SPLENDID STEAMER BALTIMORE
### WILL MAKE AN

# Excursion to Old Point and Norfolk,
# ON FRIDAY NEXT,

August 3d; leaving her berth, foot of Sixth street, at 10 o'clock A. M., arriving at Old Point at 4 A. M. and Norfolk at 5. A. M. After which the Baltimore will lay alongside the Great Eastern, and remain sufficiently long to allow passengers to board and inspect her.

Returning, will leave Norfolk at 6 P. M. Saturday evening and arrive in Washington on Sunday at 11 A. M. Passengers will be taken off and landed at all the regular landings on the Potomac.

## FARE FOR ROUND TRIP, INCLUDING MEALS, 6 DOLLARS.

For further information apply to Capt. C. E. Mitchell, on board, or at the Company's office, corner Sixth street and Pennsylvania Avenue, under the National Hotel.

### GEORGE E. MATTINGLY,
General Ticket Agent Potomac Steamboat Company.

SAGE'S Oscillating Power Press, corner of Seventh street and the Avenue.

A typical broadside showing how advertisements attracted tourists to the newest in water transport in the 1860s. For six dollars Washingtonians could steam down the Potomac to Norfolk, Virginia, and back, meals included.
Courtesy the Library of Congress.

And in 1783 the Maryland Assembly gave the port wardens power to enact regulations to prevent injury to the harbor from construction land, earth, or soil contiguous to the basin or harbor that might fill it up or obstruct it.[48] A century later the assembly prohibited disposal of ballast or earth in the Bay itself.[49]

The main problem centered in Baltimore where citizens worried about silting from at least the 1780s and thereafter. Cutting down on Hughes Street at the foot of Federal Hill caused continuous erosion into the Harbor.[50] As construction spread northwards along Jones Falls, the mouth of that stream built up alluvium in shoals.

From 1800 on, regular dredging was done in the Potomac at Georgetown, Washington, and Alexandria. By 1863 soil pollution posed a general problem along the Potomac.[51] Today twenty-five to thirty percent of the million tons of sediment entering the Potomac come from construction sites in the District of Columbia area. National Airport is built on dredged channel sediment.[52]

The midnineteenth century produced a clutch of cases concerning efforts to improve navigation. In *Garitee v. Baltimore*, the city deposited dredge spoil in front of the plaintiff's property; the court granted the plaintiff relief.[53] In *Baltimore & Ohio R. R. v. Chase*, a dispute arose between neighbors as to access rights; the court preferred the rights of the railroad which had already constructed an improvement to navigation.[54] In *Norfolk v. Cooke*, the city was granted relief which precluded a private party from filling in the tidal basin.[55]

In 1890 the national government also began regulating navigation. The United States Army Corps of Engineers established lines in various harbors beyond which piers or deposits could not legally be placed. The Corps also worked at dredging and constructing improvements like the channel to Baltimore. More will be said about their projects when we come to the proposal to enlarge the Chesapeake and Delaware Canal. This canal, opened in 1829, became the most enduring of the series of canals built in that era. The Chesapeake and Potomac Canal and the ones on Susquehanna River lost out to railroads in the rivalry for freight and passengers. We should add that Pennsylvanians in particular seem to have been keen to develop navigation. Investments by the state as well as private entrepreneurs went to building shipyards, steamboats, and hundreds of miles of canals on the Susquehanna and tributary rivers. Some Pennsylvanians even advocated blasting and dredging to make a deep channel in the shallow, obstacle-ridden Susquehanna River.

## SEDIMENTATION AT SELECTED HISTORIC TOWNS, UPPER CHESAPEAKE BAY REGION

| Town or Location | Founded | River or Creek | Approximate Time Sedimentation Recorded | Amount of Downstream Migration of Head Navigation (Miles) | Approximate Reduction In Depth (Feet) | Years |
|---|---|---|---|---|---|---|
| Bladensburg, Md. | 1742 | Anacostia | 1875 | 2 | 3 | 1875-1937 |
| Piscataway, Md. | 1634 | Piscataway | 1807 | 1 | 3 | 1863-1945 |
| Georgetown, Washington, D.C. | 1751 | Potomac | 1804 | 20 (No Dredge) | 9.25 | 1783-1837 |
| Mt. Vernon | 1752 | Potomac | 1793 | — | 1 to 4 | 1863-1904 |
| Dumfries, Va. | 1748 | Quantico | 1787 | 1.7 | 4 | 1796-1905 |
| Iron Factory | 1734 | Neabsco | — | 0.75 | — | 1734-1872 |
| Port Tobacco | 1658 | Port Tobacco | 1700 | 1 | 6 | 1800-1882 |
| Upper Marlboro | 1706 | Patuxent | 1733 | 8 | 7 | 1859-1944 |
| Elk Ridge near Baltimore | 1650 | Patapsco | Before 1898 | 7 | 15 at Hanover St. | 1845-1924 |
| Joppa Town | 1707 | Gunpowder | 1750 | 2.5 | 10 | 1750-1897 |

Professor Gordon Wolman of Johns Hopkins University compiled this table for his study in the Governor's Conference on Chesapeake Bay, 1967.

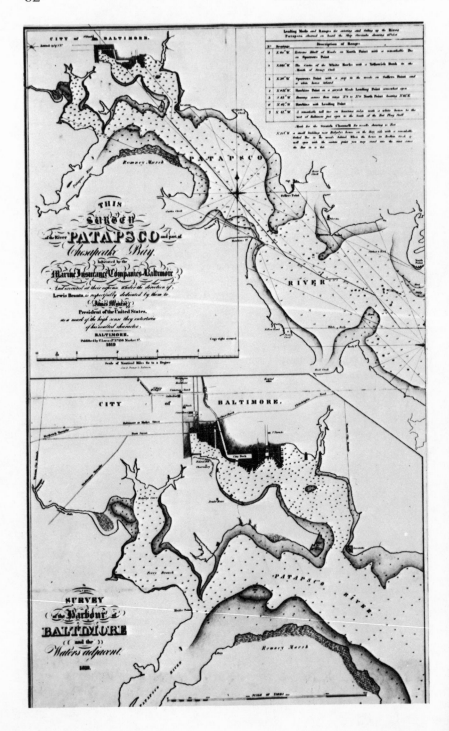

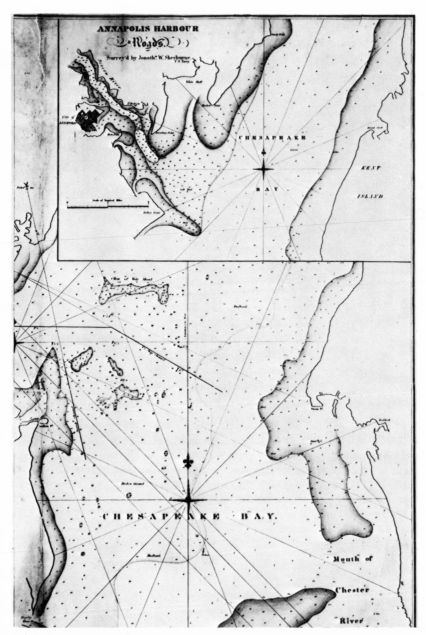

Survey of the Patapsco River and part of Chesapeake Bay by Lewis Brantz in 1819. Published in Baltimore by Fielding Lucas, Jr., the charts provided information that was essential to the growth of the port of Baltimore in the nineteenth and twentieth centuries. Courtesy Maryland Historical Society.

Richmond from the Hill above the Waterworks. Richmond, Virginia's capital, has produced a body of law governing uses of Chesapeake waters. This view of 1834 shows the James River and its canal. Through canals like this one, products of the piedmont reached ports on the fall line such as Baltimore and Richmond. These idealized views made Chesapeake towns seem both handsome and prosperous. Courtesy the New York Public Library.

By the end of the first three hundred years of the white man's use of the Bay, another improvement that came under the eye of the federal government was erecting lighthouses. The United States Coast Guard took a role only long after the individual states had erected most of the seventy-four that were built. The states began building when steamboats arrived. In the early times, lighthouses stood like the bobbins of a knitting factory, guiding the skein of steamship lines that stretched all over the Bay.

The lights also guided fishermen as they "traveled the dark." A number of these lights still function, and four of them were still manned as late as 1970. Others, abandoned, stand recycled as houses, or in ruins, or else they do not stand at all. Once the Bay's shores held more screw pile type lighthouses than any other place. One curious historical fact: though the Chesapeake had the greatest volume of shipping in eighteenth century America, it was the last important region to get what one man called "noted landmarks to guide the doubting mariner."[56]

Lighthouses, however, made only slight territories for the federal presence. More significant, the seat of the federal government itself faced the Potomac. Congress had also taken over more and more of Annapolis for the United States Naval Academy since the school moved there in 1847. Then too, the Navy Department created a major base for its ships in Hampton Roads at Gosport after the Civil War. In 1917 the deep water harbor there became the greatest naval base in the country.

This record size points up the shift from fish to ships in the first three hundred years of the Bay's history. This naval base also has significance because it illustrates the twentieth century dominance by federal presence. But much more of that story will follow in succeeding chapters.

# SIX

# Land of Pleasant Living

Chesapeake Bay, land of pleasant living.
　　　　　—a Baltimore beer commercial (mid-
　　　　　twentieth century)

We must remember that boats were sailed for pleasure as well as commerce. For over two centuries men and women skimmed over the surface of Chesapeake waters, merry and free. Joseph Seth, for example, recalled in 1926 the sailing canoes of the 1870s: they were swift sailors—never surpassed, he wrote, for beauty of line. "The days spent in our 'Flying Fish' were some of the happiest of my life." He noted that even the small canoes were equipped with sails, that almost every family possessed one, and that this kind of boat had played an important part in all regattas since colonial times. Of sailing parties by moonlight, Seth remembered, "The union of the beauty of the night and the graceful flying canoe were the essence of poetry and romance."[1]

Since the seventeenth century, private boats like the "Flying Fish" have been sailing pleasurably across uncrowded ways. In 1689, for example, Mayor Robert Sewall and his ketch *Susanna* provided a footnote to recreation on the Bay. Then in 1760 we have record of an organized regatta. Later in that century Edward Lloyd of Wye on the Eastern Shore wrote his London agent:

> Be pleased to send me a complete set of American colors, for a pleasure-boat of about 60 tons burthen and six brass guns to act in such manner as to give the greatest report, with the letters E. Ll. thereon.[2]

In 1825, the Baltimore *American* advertised pleasure boats for hire on Fells Point, Baltimore. No wonder artists painted idyllic waterscapes with backdrop of domes and steeples that make that city look like

Constantinople. Even back then, the Inner Harbor held more than just commerce.

Around 1800, lower down the Bay the schooner *Dolphin* carried "on board a number of Gentlemen on a party of pleasure." The reporter added that, equipped with a seine and a variety of fishing tackle, they were "abundantly supplied with fish . . . and to crown their felicity, they found no vessel in the bay able to sail with them [that is, capable of outsailing them]."[3]

Such yacht racing was not for everyone. But luckily, the masses could also make "parties of pleasure" on the Bay, once excursion steamboats began plying the waters after the Civil War. In summer, people flocked aboard to cool off. Some passengers spent the day at resorts like Chesapeake Beach, Calvert County, Maryland, or on the Eastern Shore at Tolchester or Betterton. In its brochure, Chesapeake Beach offered "Finest Salt Water Bathing," fishing, crabbing, yachting, and "Attractions not offered by any other Resort this side of Atlantic City."

Other escapees from August heat waves danced the evenings away as they cruised on the *Emma Giles* or sister ships. How many people cruised we can learn from the account of an accident. Fourteen hundred excursionists aboard the *Louise* were returning to Baltimore from Tolchester on July 28, 1890. About 8:15 P.M. as they were nearing Seven Foot Knoll, their ship was hit by the *Virginia*, a steamer then on its regular run from Baltimore to Norfolk. Fourteen people were killed.

Seventy years earlier with the coming of the steamboat, Virginia excursionists steamed to bayside resorts too. Some people went there with special purpose, to sustain or regain health. As early as 1854 Ocean View, Virginia, attracted invalids because it offered quiet and recreation. Any invalid or convalescent who was up to exercising could have walked on the beach, ridden horseback, hunted ducks, or fished. The visitor could simply have basked in the salubrious air, as no doubt health seekers did a generation earlier at Old Point in Hampton Roads. In 1821 a hotel opened there, appropriately called Hygeia. Health seekers merely opened the way. Soon they were outnumbered by "public men, weary of their cares, army and navy officers furloughed or retired, and the gay daughters of Virginia."[4]

Through all the later decades of the nineteenth century, recreation abounded for both invalids and noninvalids. Who wouldn't have found pleasure in this region of clean estuaries and rich marshes? A

The clean, white excursion steamboat contrasts dramatically with the smoke
darkened horizon of Baltimore. Belching smoke on the left is from the power
plant that ran streetcars for years. In 1905 when this photograph was taken,
excursionists returning from Betterton across the Bay always commented on
the fact that they knew they were within five miles of Baltimore by the putrid
harbor smells. Courtesy the Edward L. Bafford Photography Collection, Albin
O. Kuhn Library and Gallery, University of Maryland, Baltimore County.

sky full of ducks and geese brought hunters. By 1814, we know, hunters were using "wooden figures cut and painted to represent ducks."[5] Then to meet another need, clever breeders mated Newfoundlands with water spaniels to produce the Chesapeake Bay Retriever.[6] Down the years human ingenuity in devising decoys, dogs, and blinds balanced human laws such as bag limits and set seasons— all to the furtherance of man's pursuit of a duck.

Hunting stories take second place in glory only to fishing tales. Nostalgia infuses both. Remembering 1850, Joseph B. Seth wrote:

> My mother was a great lover of fishing. . . . It was her almost daily custom in August and September to paddle out on the water in her special rowboat about 4 o'clock in the afternoon, stopping to throw her lines not more than twenty-five yards from the foot of the lawn. Here she would fish for two hours. Her luck was proverbial, as she rarely failed to bring home a good catch. These fish were often used for our supper, literally almost going into the pan alive and kicking. Good fish eaten under these conditions made delectable food.[7]

The same enthusiasm can be demonstrated for crabs, though it developed later. In 1836 a Philadelphia doctor reported that "Soft crabs are, with great propriety, regarded as an exquisite treat by those who are fond of such eating," although he noted that many people wouldn't touch either lobsters or crabs. He added, "There are few who taste of the soft crabs without being willing to recur to them."

This Philadelphia physician, Dr. John D. Godman, called softshell crabs "an article of luxury." He continued:

> They are scarcely known north of the Chesapeake, though there is nothing to prevent them from being used to considerable extent in Philadelphia, especially since the opening of the Chesapeake and Delaware Canal [1829].

He provided other information that may startle twentieth century crab eaters:

> The price at which they are sold is sufficient to awaken all the cupidity of crabbers. Two dollars a dozen is by no means an uncommon price for them when the season first comes on. . . . The slaves search for them at night, and then are obliged to kindle a fire of pine-knots on the bow of the boat, which strongly illuminates the surrounding water, and enables them to discover the crabs.[8]

A more modern scholar, L. Eugene Cronin of Chesapeake Bay Laboratory, asserted that crabs positively had been eaten near the Chesapeake since 1855, when crabbing was the basis for a small industry. Crisfield entrepreneurs shipped crabs from the 1870s on. Back in the 1730s William Byrd ate "flat crabs" about as large as a hand.

Pleasure in swimming in Bay waters must have peaked around the turn of the twentieth century, or even later. This recreation did not last. Indeed the custom of bathing from beaches had begun for the masses only late in the nineteenth century. Earlier of course it would have been hard to keep boys out of the water. Now a dignity far above the Tom Sawyer level of skinny-dipping settled over resorts like Virginia Beach, Virginia, and Betterton, Maryland. There ladies in black stockings and skirted bathing dresses disported on the sand, as they appeared at the opening of the first municipal bathing beach in Baltimore, a stretch of Canton shore, just east of Fells Point, Baltimore, in 1893. This spa was fitted out with floats, ropes, and crude cabins for changing. Suits could be rented. So popular was recreating in the rather polluted Patapsco water that the number of bathers during the second season was 27,787.[9]

At that time and just east of this beach, Back River neared the end of its long usefulness for recreation. Sportsmen had loved hunting and fishing there. Then Baltimore City authorities chose 547 acres at the western end of the Eastern Avenue Bridge for the sewage treatment plant.[10] (The significance of this plant for other aspects will be discussed in Chapter 7. For our discussion here, however, it meant the end of pleasure for sportsmen at this site.)

Still, other Bay waters and marshes filled the need for a time, though more distant from the metropolis. It is important to recognize that the compactness given by rowhouses had always given Baltimoreans easy access to the country. In the early twentieth century, however, the city began to sprawl. One last chance to preserve nearby land and water came in 1904 when the famous Olmstead Brothers' firm of planners came to town, hired by the Municipal Arts Society of Baltimore (a private group). They came fresh from having created a chain of parks for Bostonians, including Riverside near the mouth of the Charles River. In Baltimore, the planners expressed dismay that no public park opened the Bay to citizens.

They found the perfect spot for just such a water park only seven miles south of the center of town. They chose Curtis Creek and its

Fortress Monroe, Old Point Comfort, and Hygeia Hotel, Virginia, in 1861 and 1862. The Key to the South. In the 1840s vacationers and invalids escaped the cities, and in places like the Hygeia Hotel, with its stately columns, found healthy recreation on the Bay. Courtesy the Library of Congress.

Popular recreation at waterside resorts around the Bay included swimming, sunning, and sometimes an opportunity to ride on a Ferris wheel or a merry-go-round. Pictured here is a typical beach at Arlington, Virginia, in 1925. Many beaches were later closed to swimming. Courtesy the Library of Congress.

tributary Furnace Creek. There, the team wrote, eight hundred acres enclosed eight hundred acres of water to create an area ideal for small boating, for picnicking, and for swimming and water sports. This new park would have given Baltimoreans access to Curtis Bay, to broad and quiet tidal estuaries, to "scenery which forms the most beautiful and characteristic landscapes of all the country lying along Chesapeake Bay."[11]

Nothing came of this proposal. So Bay lovers lost an amenity. The Olmsteds' genius lost a monument. And Baltimoreans lost a nearby escape from crowded, hot streets on Chesapeake waters.

Certainly town dwellers had to escape then just as they do now. Tidewater towns, it would seem, have never been agreeable places to live. Hampton, Virginia, for example, stands as the oldest settlement of English America in continuous existence, but its marshes and mud banks early made it an unhealthy spot to live in. A visitor in 1796, Isaac Weld, reported that Hampton was "a dirty disagreeable place, always infested by a shocking stench from a muddy shore when the tide is out."[12]

Again, about Norfolk, Virginia, William Byrd II wrote, "With all these conveniences [for trade], it lies under the two great disadvantages that most of the towns in Holland do by having neither good air nor good water."[13] Other locations around the Bay, happily, did possess the air and water deemed healthy by eighteenth-century colonists. Both Richmond and Baltimore benefited from sites on the Fall Line, with hills and fast running streams.

But Norfolk suffered another drawback. Visiting there in 1794, Mederic Moreau de St. Mery complained about the new houses springing up in the direction of the Elizabeth River and of the new wharves built in haphazard arrangement "put up solely for the convenience of the owner and built without any general plan, and inconsiderately shut off the view of the river."[14] Trust a Frenchman's eye for the esthetic to notice what was wrong with Bay building. Waterfronts filled with warehouses. There men like Byrd exploited the waters without a thought to esthetics (or water quality).

Away from towns, this French visitor might not have found fault. Being domiciled above, near, and on the water satisfied many a man or woman's esthetic sense. Recollecting Eastern Shore life in the midnineteenth century, Joseph B. Seth (born in 1846) wrote:

No pleasanter memories arise in my mind than those associated with the beautiful sheet of water upon which our home was situated. The water was salty, it being tributary to Chesapeake Bay, and flowed in and out twice a day. It was not a rapid tide, but amply sufficient to keep the water pure.[15]

About the time Seth's great-grandfather surveyed this beautiful property in 1780, George Washington was appreciating the location of his house on another Chesapeake site: "No estate in United America is more pleasantly situated" than his Mount Vernon.[16] Contemporary painters always showed the mansion crowning its plateau above the wide Potomac and with ships of trade sailing past. (See page 27.) Similarly placed above tidal waters, the State House and Governor's Mansion in Annapolis and the Capitol in Richmond invited people on shipboard to admire how architecture ordered the wilderness. (See page 64.)

Lacking such grand vistas of water, English kings and dukes constructed artificial canals and lakes like those at Hampton Court and Blenheim. More fortunate, Chesapeake planters created grander prospects near Annapolis, such as Java (on the Rhode River) and Tulip Hill (on the West River). Builders at Mulberry Fields in southern Maryland carefully framed the water. To make the river, a mile away, seem nearer, they created an illusion in perspective. They planted the avenue of trees flanking the view to fan out from the house to the water.

The special appeal of views of the Chesapeake is shown in these real estate advertisements from the *Baltimore Gazette & Daily Advertiser:*

> For Sale or Rent A very valuable Farm in Harford County, beautifully situated, commanding a fine view of the Chesapeake Bay and Gunpowder River . . . a never failing spring of excellent water at a convenient distance, and a well within a few steps of the door . . . 500 acres, 17 miles from Baltimore . . . Col. Edward A. Howard          —Tuesday evening, August 26, 1828

---

> To rent or lease for a term of years. The Country Residence of the late Alexander H. Boyd, not more than 20 minutes walk from the centre of the city . . . The view of the River, Bay and surrounding country is not surpassed by any residence near the city.                            —Friday evening, February 29, 1828

No matter how agreeable the amenity of a view, colonists had a practical reason for choosing the site—health. To them, a healthy site meant a hilltop with good air. The Lee family's mansion Stratford, for example, rises some distance from the Potomac River. The builders, however, did erect two platforms within groups of huge chimneys, vantage points from which we can still follow the band of water to the horizons. Though the unusual chimneys resemble designs by Sir John Vanbrugh for Claremont in England,[17] the view belongs to the Chesapeake.

Occasionally, though, vistas and health gave way to placing house and office near warehouses and shipping. Sited close to the James River, Westover (1730) was headquarters for William Byrd's two hundred-square-mile empire. Byrd and the other planters tell us a lot about attitudes towards governing Chesapeake waters. These eighteenth century men had to accept change. And, in their isolation, they had to live with independence from government control. The uncluttered landscape and waters of course imposed their own demands for adaptation. Even the architecture changed in the new climate. Curiously, no single English house served as source for Virginia's mansions, though prototypes may include Coleshill, Berkshire (designed probably by Roger Pratt about 1650) and Thorpe Hall, Norfolk (Peter Mills, architect, 1654-1656).[18]

Today we envy planters like the Byrds and the Lees. We wish we had their amenities—uncrowded space, clean air and water, the slow pace of life. Few of us in the twentieth century can spend the afternoon the way this Virginian and his wife did in 1759 and 1760:

> September 20: Fine weather. Went in the afternoon and drew the seine. Had very agreeable diversion and got great plenty of fine fish.

His diary records other good times:

> September 26: Went with my wife in the evening to draw the seine. Got sixty green fish [black bass] and a few other sorts.
> October 6: Went with my wife to see the seine drawn. We dined very agreeably on a point on fish and oysters.
> January 22, 1760: Bought about 70 gallons of rum. Got fine oysters here.
> February 9: Went with my wife and Mr. Crisell to draw the seine. We met in Eyck's Creek a school of rock, - brought up 260. Some very large; the finest haul I ever saw. Sent many of them to our neighbors.
> February 22: Drew the seine and got 125 fine rock and some shad.[19]

# An Immense Protein Factory

Baltimore lay very near the immense protein factory of
Chesapeake Bay, and out of the Bay it ate divinely.
—H. L. Mencken (1940)

We have discussed the development of the Bay region over its first
three hundred years, and described the origins of various uses of the
Bay and its resources. By 1900 the dominant Bay uses of waste
disposal, navigation, and fisheries were well established, and there
was considerable recreational activity as well. Conflicts among these
uses had been few. But four important forces had developed which
would directly influence the Bay. Foremost was the development of
the cities of Hampton Roads, Richmond, Washington, and Baltimore.
With them came the need for water supply and municipal sanitation.
Concurrently developing was the large scale use of the Bay as a waste
placement sink. This use was related both to the growth of cities and
their water supplies, and to the growth of industry. Deepwater navigation by merchant ships was also on the rise. Last in economic
importance, but perhaps first in terms of political attention, was the
growth in commercial fishing, most particularly the oyster fleet.

Commercial fishing on the Bay prior to the nineteenth century
was primarily for local consumption and attracted relatively little
political attention. By the early part of the nineteenth century, New
England oystermen were making raids on the abundant oyster bars of
the Chesapeake. Not only were the New Englanders taking oysters to
northern markets centered in New York and Philadelphia, but they
were also using Bay seed oysters to restock their bars, which had been
stripped by earlier generations of New England fishermen. Both Virginia and Maryland moved to prevent out-of-staters from capitalizing
on the Bay's resources, and there gradually developed a robust Bay
oyster industry, which peaked in the 1880s.[1] Despite exclusionary

legislation, much of the impetus for this development still came from New Englanders, who financed many of the largest harvesting, processing, and shipping operations.

Development of the oyster industry has been a dominant factor in the evolution of Bay water-quality policy and politics for nearly a century. Virtually all significant issues over quality have had the welfare of the oyster and the oysterman as a central concern. It is, therefore, useful to outline the development of the industry, in order to explain its central importance.

Oysters in Chesapeake Bay are nearly ubiquitous except in the upper reaches of the Bay and its tributaries, areas of low salinity. They occur in commercial quantities on beds variously known as bars, reefs, rocks, or knolls. These bars are potentially self-sustaining. Oyster larvae spend a period of time as a free-floating part of the Bay zooplankton but eventually must settle on a hard surface to begin the development of a shell and their growth to maturity. An ideal surface is the shell of another oyster, living or dead. Thus over time, a favorable site will develop an expanding area and quantity of oysters, built on the shells of oysters that preceded them. In some places these bars grew to within a few feet of the surface, hence the name "reef." It was these natural bars that attracted the New England oystermen and later provided the crop that reportedly totaled over twenty million bushels a year during the 1880s.[2]

Bars built by nature can also be destroyed by natural means. Long-term freshening of the upper Bay and the Potomac have eliminated once productive beds leaving vast quantities of shell. The constant shifting of the Bay bottom and erosion of its shoreline have silted over bars that could not grow upward fast enough to keep ahead of the deposits. And natural enemies such as the starfish, the moray eel, or various smaller predators, and diseases, could, and no doubt did, eliminate bars. But in the nineteenth century, and by general reputation still in the twentieth, the Bay was perhaps the most productive natural oyster ground on earth.

Man also can both destroy and build oyster beds, in the latter case requiring assistance of nature. Destruction can be indirect through the modification of salinity patterns in the Bay or by increasing sedimentation and erosion.

But the most dramatic destruction of oyster beds comes from overfishing. Overfishing made large areas nonproductive during the heyday of the oyster raids of the late nineteenth century. Despite

Sailing log canoes provide some of the best pleasure sailing in the country. The well-known photographer A. Aubrey Bodine captured a scene on Virginia waters the way people in the middle of the twentieth century liked to remember the Bay. Courtesy A. Aubrey Bodine Collection, the Peale Museum.

growing concern for the decline of the industry, and a large body of law to prevent it, beds were stripped of their standing crops of mature oysters. In some cases, beds were rendered unfit for future productivity.

On the positive side, man can establish a bed for oyster production, using techniques perfected in France in the early part of the nineteenth century. Oyster shells can be dumped in an otherwise fallow area, and either receive a natural set of spat, or be provided with seed oysters from another area. Through a process of preparing the ground, planting, thinning, and preventing excessive siltation and predators, man can produce oysters in a fashion analogous to land farming. The same techniques can increase the productivity of natural bars.

Government involvement in the oyster industry was, as mentioned, extensive in the nineteenth century. The "oyster question" occupied much of the time of both state legislatures. The first task was to prevent outsiders from stripping the bounty of the Bay. The second was to prevent residents from overfishing. The third was to resolve the various disputes between different counties, since most regulation was done on a county-by-county basis. The fourth was to address differences between hand tongers, who generally worked the more protected and shallow waters of the Bay's tributaries, and the dredgers, who worked the larger bars and were regarded as the chief sources of oyster bar destruction.

Last, but not least, legislatures had to address the question of whether to allow the leasing of Bay bottom for the farming of oysters. Clearly it would not be profitable to go to the expense of developing a bed if someone else could harvest it. Many observers of the great oyster boom believed that the future of the industry lay in private oyster culture. But the oystermen who made their living harvesting naturally grown oysters did not want to be denied that opportunity.

Virginia regulated a compromise. The solution was to establish that all natural bars, as delineated in a late nineteenth century survey, were to remain as common property resources, open to harvest by anyone who met residency requirements and who observed laws relating to licensing, gear, and season. At the same time, a method was established to lease fallow bottom to individuals or corporations, thus permitting the development of a substantial private culture as had been developed in Europe, New York, and New England.

The photograph creates a nostalgic quality in the workaday world of commercial fishermen. In the foreground are sixty-foot weir poles pulled up on shore at end of season, Horn Harbor, northeast of Mobjack Bay, Virginia. Courtesy A. Aubrey Bodine Collection, the Peale Museum.

In Maryland, the oyster was seen to be a prime economic asset.[3] Since 1884, its fisheries officials had argued that the Bay was a tremendously productive body of water, capable of producing far more oysters than occurred naturally. The institution of oyster farming, following the methods developed in France, could make the yields of the Bay greater than those even of the heyday of the great raid on the natural bars during the 1880s. (Indeed some officials at that time argued that the faster the natural beds were depleted, the sooner the state could turn to the proper management of the resource, allowing for a stable and economically important industry comparable to dry land farming.)[4]

For complex and varied reasons, the concept of private oyster farming was never accepted widely in Maryland. Factionalism between watermen from different counties, between dredgers and tongers, and between all watermen and the conservation agencies, drained whatever energy there might have otherwise been to get about the business of managing the oyster more efficiently. The only thing the watermen agreed upon was their opposition to a private oyster culture.

Maryland oystermen formed the nucleus of what was to become a powerful political force. Since oysters had to be shucked, canned, packed, and shipped, subsidiary industries employing thousands of workers had sprung up throughout the region.[5] The economic atlas for the state of Maryland at the turn of the century illustrates not only the numbers of persons involved in the industry, but also the fact that they were distributed throughout the tidewater counties.[6] Because each county then had a single senator and also because the tidewater counties outnumbered their nontidal counterparts, political representation for the oyster interests was disproportionately large.[7] By the turn of the twentieth century the Maryland oyster lobby had political clout greatly in excess of the economic importance of the oyster industry.

The commercial harvest of finfish has never gained anything like the size or political significance of the oyster industry in the Bay states. It was a matter of considerable attention in the nineteenth century, however, particularly the shad harvest. Shad runs had apparently long supplied a substantial portion of the food of some of the poorer Bay area residents, and reductions in spring runs had been lamented at least since the 1920s. The numerous statutes relating to the damming and blocking of streams were aimed partly at reducing the obstructions to the spawning of shad. Declines in supply had the effect of driving up prices, a rise that was decried as having an unfortunate impact on the poor.[8]

Declines of finfish through the eastern United States had much to do with the formation of the Federal Fisheries Service. This agency in turn led to the establishment of state fisheries boards in both states soon after the Civil War. Along with the oyster-police force also established during this period, this constituted the first steps to creating a state bureaucracy for the management of Bay resources. Although these boards were small in 1900, they provided a permanent government presence that represented the interests of both sport and commercial fishermen.

Hence development of the commercial fishery provided the Bay with a twentieth century legacy. Maryland and Virginia had gone their separate ways in terms of fishery management. And they were to continue to squabble. Some of the most heated and involved disputes were over the location of state boundaries in the vicinity of Tangier Island and along the Potomac River. Another hardy perennial dispute was the debate between Maryland and Virginia over the latter's practice of allowing the winter harvest of female crabs.[9] There were also frequent discussions about the effects of the harvests of finfish on the supplies of both commercial and sport species. Maryland, for example, imposed a prohibition on the purse seining of menhaden, arguing that menhaden were the bait fish that attracted large numbers of more desirable predator fish to the upper Bay. Virginia, on the other hand, allowed this efficient method of harvest of menhaden, and a thriving menhaden fishery was established in the lower Bay.[10]

Criticism of Maryland's public oyster fishery continued into the twentieth century. A 1933 Conference restated the allegations of mismanagement of Maryland's oyster resources. An official stated that pollution was a "scapegoat" for overfishing. The Baltimore papers, in numerous articles, chided the Maryland General Assembly for failing to capitalize on the productive potential of the Bay. In one particularly pointed article, the oyster industry was described as being on an economic par with "condensed milk and window shades," rather than being the source of vast wealth that was its heritage and its potential.[11] But the Maryland oyster lobby was politically entrenched. It would continue to be successful in its efforts to rebuff these criticisms; it also stood ready to fight new battles in support of the welfare of the oyster industry, if need be.

A lone crabber in white boat, August, 1982, near Calvert Cliffs on Chesapeake Bay. In winter this waterman would probably hand tong for oysters and fish in spring and fall. Photograph by Richard McLean.

Finally, federal intercession had resulted in creation of the Bay's first bureaucracy. The fishery boards which had been created in Maryland and Virginia had but few powers, but their successors were to play an important role in addressing Bay pollution.

# EIGHT

# Sewage and Shellfish

> One of the greatest questions for the future is that of
> pollution. The pollution of sewage of our ever increasing
> population and the waste from our rapidly growing in-
> dustries is affecting the entire fish and oyster industry in
> and around Hampton Roads. . . .
>
> —Annual Report, Commissioners of Fish-
> eries of Virginia, 1918-1919

As the cities of the Bay region grew, they had to deal with the twin problems of water supply and municipal sanitation. Springs and small streams gradually proved inadequate to supply water for the growing populations, which by the latter part of the nineteenth century began to seek the advantages of indoor plumbing. Now the Bay towns, like cities elsewhere, formed municipal water companies to distribute the water through a central system. Richmond and Washington were able to use the large flow of the James and Potomac rivers, while Baltimore reached out into the surrounding country to impound small rural streams. The Norfolk area was forced to rely almost entirely on wells. By the 1870s, all the Bay's cities were using substantially more water per capita, not only for domestic use but for street cleaning, fire protection, and industry.[1] This water, laden with wastes, quickly reached the Bay. Because of the relatively sluggish tidal exchange rates of the Patapsco basin in Baltimore, the upper Potomac estuary in Washington, and the small tidewater creeks off Hampton Roads, all these waters were directly offensive to the senses, at least at certain times of the year by the end of the Civil War.[2]

Concurrent with municipal growth was the concern for control-ling contagious or epidemic diseases. Prior to the last decade of the nineteenth century, the prevailing theory was that miasmas, or foul air, transmitted most of the diseases then highly feared, such as cholera, dysentery, and yellow fever. Special attention was, therefore, given to municipal drainage, to the elimination of swamps, and to the design of sewers, so that the latter would not result in the buildup of sewer gas. Contaminated water had, of course, also long been recognized as a source of disease, although the causal agent had not been determined until the germ theory of disease was established just before the turn of the century.[3]

As the result of this interest in water supply, wastewater removal, and public health, the cities around the Bay formed public works bureaucracies starting with municipal water companies, and then going to sewerage commissions.[4] At about the same time, local and state agencies were formed to study public health issues, apply quarantines, and police sanitary practices.

## THE BALTIMORE CASE

By 1900 the condition of the Baltimore Basin was a longtime topic of agitated debate. In 1893, a sewerage commission was appointed to tackle the problem afresh, since the reports of earlier commissions (1862 and 1883)[5] had not led to action. The commission dutifully set out to study the various alternatives, which it did in great detail, making use of some of the leading authorities of the time. In 1897, it submitted its report,[6] which recommended the building of a sewer that would discharge the untreated wastes of some 350,000 people into the Bay. This approach, which would be unthinkable today, was labeled as the "water carriage-dilution method" of waste disposal, a rather euphemistic term for what another writer vividly called the "construction of a great artificial intestine and anus" for the city.[7]

Had the commission done its careful work just a few years before, its recommendation would likely have been acted on. However, in October 1893, an event took place in Connecticut that directly and dramatically affected the outcome of the Baltimore sewer issue, and established what was to become the dominant Bay water quality issue of the next fifty years. A group of Wesleyan University students became ill with typhoid fever after having eaten oysters.[8] Medical science, armed with the newly accepted germ theory of disease, firmly established that the oysters were the direct cause of the disease. The incident received worldwide medical and popular attention and sent a shock wave through the Bay oyster industry. For the first time, what had been a mere theory, scorned by many, that oysters could cause disease, was now a proven and widely known fact.[9]

No wonder then that the oyster industry, represented both by watermen and packers, voiced forceful objections to the discharge of Baltimore sewage to the Bay.[10] Thus, when the sewerage commission submitted its 1897 report, its entire argument was designed to offset the objections of the oyster interests, and to demonstrate that the proposed discharge to the Bay would not cause harm. The commission pointed out that the only incidents of disease came from oysters taken from confined and poorly circulating waters, whereas the proposed discharge would be rapidly diluted and circulated by the immense volume of water in the middle of the Bay. It argued that the cities along the Susquehanna were discharging the wastes of over one million people to the Bay, without ill effect. Furthermore, the oyster beds closest to the proposed outfall had been seriously depleted or eliminated by floods of fresh water from the Susquehanna, hence reducing even further any chance of contamination. Indeed, the Bay was productive because of the runoff from the land. The sewerage commission stated:

> Your Commission has been advised by an acknowledged authority on all that related to the biology of the oyster that a discharge of the sewage of the city, as here contemplated, would be beneficial rather than injurious to the oysters themselves.[11]

Speaking with the assurance born of careful deliberation and sweet reason, the commission concluded:

> There would appear to be but little reason why the City of Baltimore should deny itself the facilities and advantages which nature has vouchsafed to it namely the diluting effect of the Bay, since the Susquehanna and other towns are doing it anyway. It would be ridiculous to attempt to prevent it. . . . May not Baltimore, as well, without offense to others, purify herself in the broad waters of this great bay without thereby disturbing or annoying any existing interest? Your Commission thinks she may.[12]

Other, and more potent, interests did not. Although they apparently did not resort to rhetoric, and, therefore, have left us no written record of their argument, the oyster factions prevailed on the mayor and council to reject the commission's proposal. The local medical community was also involved, although apparently of split opinion, one group arguing that there was no medical risk, the other arguing that pathogens could survive for long periods of time in both salt water and oysters. And finally, there were the objections of a small but well-organized interest group, the night soil contractors, who made their living collecting the wastes from vault toilets and selling it to area farmers.[13]

These forces combined to persuade the mayor and council to reject the proposal of the commission. In his 1897 message to the city council, the mayor pointed out that the city had a strong economic stake in the oyster industry, and, therefore, should not approve a discharge to the Bay.[14] He urged the adoption of "the land filtration technique," an option discussed by the commission but rejected because of its high cost relative to direct discharge to the Bay. This proposal called for pumping the sewage to the Glen Burnie area in Anne Arundel County, a few miles to the south of the city. There the wastes would be used to irrigate and fertilize farmland, a practice widely used in Europe since about the middle of the century. The mayor urged prompt action, exclaiming of the project, "it must be done!"[15] The council agreed to disagree with the commission, and, in 1898, the mayor and council issued a joint resolution rejecting direct Bay discharge.[16]

The commission obligingly went back to work and issued a second report in 1899.[17] When the same members met, they clearly had not changed their opinions as to the soundness of their original recommendation: they spent some time in bolstering the arguments for direct Bay discharge. To their previous list of arguments they added the observation that much waste from Baltimore reached the Bay anyway: through exchange between the waters of the Patapsco and the Bay; through the disposal into the Bay of material dredged from the Baltimore harbor (an observation that presages one of the major issues of the 1960s and beyond); and, from runoff from farms that used Baltimore City night soil. Furthermore, argued the commission, if people wanted to worry about the sanitary conditions of oysters, they should worry about oysters harvested from the shallow waters near towns on the tributaries of the Bay. These towns, although not named, surely included Cambridge, Crisfield, and Annapolis, all of which had sizable watermen populations. With a combination of petulance and stiff-necked pride, they concluded their defense of the dilution approach by saying:

> However, if our fellow citizens are not inclined to avail of this the dilution method, the natural and most economical method of disposal, and should determine to have something else, leav-

ing the economies of the dilution method to be enjoyed by our neighbors, . . . nothing remains for your Commission . . . than to point out such other methods as seems to it available, though it be much more costly to construct and operate.[18]

With that final blast, they turned to the alternatives and recommended that the best approach was to treat the sewage and then discharge it to the Bay. Other alternatives combined various treatment strategies and disposal points, as well as the use of Glen Burnie land disposal for at least part of the sewage from the southern portion of the city.[19]

For political and economical reasons, the city did not act on this set of recommendations. It remained for the great Baltimore fire of 1904 to clear the way, both physically and politically, for the construction of a comprehensive sewer system. By that time, all other major cities in the country had done so. But by being last, Baltimore vaulted into the lead in sewage treatment, as we will describe. The city government quickly turned to the Maryland General Assembly to get the authorization to form a sewer authority and issue bonds for the construction of a system, an authorization required by the state constitution. The General Assembly gave its approval, but in so doing stipulated that:

> . . . said commission, to be formed under the Act, shall have no authority to construct and establish any sewage system involving the discharge of sewage, as distinguished from stormwater or ground drainage, into the Chesapeake Bay or any of its tributaries.[20]

The oyster interests, in clear control of the matter through the heavy representation of tidewater senators and delegates, thus prevented any possibility of the city making use of the Bay for direct discharge.

Faced with this limitation, the newly constituted commission set about once again to examine the alternatives and come up with a proposal. The first issue was how to interpret the restriction placed on them by the legislature. A strict reading would have suggested that no discharge, treated or untreated, was permitted. Since this was clearly impossible (except through a total land disposal system), the commission, by resolution, adopted the following approach: ". . . that the effluent proposed to be discharged into the Chesapeake Bay or its tributaries in the system to be recommended by the engineers shall be of the highest practicable *degree of purity*."[21] (Emphasis added.)

Opening in 1909, Back River Disposal Plant was part of a system which gave Baltimore the best sewerage in the nation at that time. In this photograph taken by Alfred Waldeck, October 18, 1922, sludge is being dumped into a truck. Courtesy Enoch Pratt Free Library.

The commission turned, for the fifth time, to the leading sanitary engineers of the Western world. With amazing speed and efficiency, in view of the indecisiveness of the previous forty years they adopted a plan to build a sewage treatment plant on Back River. The plant was to use sprinkling filters, settling basins, and sand filters, an unprecedented undertaking. Said one of the consultants: "No city in our country of the size of Baltimore has as yet been required to give its sewage treatment with the object of obtaining the highest practicable degree of purity."[22] He added "We are aware that the works proposed at Baltimore constitute by far the largest undertaking of sewage purification in this country." Indeed, when the plant was put into operation in 1912, it was regarded as one of the engineering wonders of the modern world.

In adopting this plan, the commission was careful to point out that "Owing to the small flow of upland water and the limited range of the tide, there are no strong currents in Back River, and the effluent so discharged would not, under usual conditions, reach its mouth for several weeks. We are of the opinion that there would be no danger from this discharge to the oyster beds of Chesapeake Bay, and no offense along the shores of Back River or elsewhere."[23] The proposal was also checked "with oyster authorities" and, approved, proceeded apace.

It is perhaps difficult for the modern reader to appreciate the unprecedented nature of this commitment to treat the sewage of a major city discharging to an estuary. Not only was Baltimore the first major city in America to adopt a waste treatment system, but also its reason was far removed from those commonly used to justify, or require, the treatment of sewage. The strongest impetus for sewage treatment was probably where such treatment would help to protect a city's own water supply. For example, Chicago both discharged its wastes and drew its drinking water from Lake Michigan. A second incentive was that a downstream neighbor drew its drinking water from the river used by another to dispose of its wastes. A third was that an acute nuisance prevailed, one that for some technical or economic reason could not be solved by piping the wastes away from the city. A fourth was that the disposal of wastes was clearly injurious to fish, recreation, or other resources. None of these reasons applied to Baltimore. Had the wastes been piped raw to the middle of the Bay, no water supplies would have been affected, no nuisance conditions created, and, most probably, no major change to the aquatic life of the

Bay induced. Yet, pathogenic bacteria might find their way from a diseased human being to the Bay, thence to an oyster, and finally to another human being. So Baltimore spent an enormous amount of money to treat its wastes, before discharging them into a sluggish tidal tributary of the Bay. As we will discuss, it was not until the late 1930s for Washington and the 1950s for Norfolk that the other two Bay metropolitan areas put sewage treatment plants on line.

The Baltimore sewer story is significant also because it gives a well-documented picture of how city officials or, more particularly, their study commissions and consultants, went about developing their recommendations. The commission reports of 1897, 1899, 1905, and 1906 reflect a remarkably efficient, flexible, and sophisticated approach to the problems. They should be required reading for anyone—professional, citizen, or politician—involved in contemporary sewage treatment controversies, such as the perennial debates involving sewage collection and treatment in the Washington metropolitan area. For they are useful models of how to bring together the best available technical assistance, evaluate the recognized alternatives, reach a reasoned conclusion, and present a written report, all in a relatively short time.

## VIRGINIA

While Maryland was going through the resolution of a major water-quality issue in the form of the Baltimore sewage controversy, Virginia was affected in a more insidious way by the aftermath of the Wesleyan incident. As the first effect, local boards of health in the Hampton Roads area began to question the safety of, and then closed to harvesting, some of the leased oyster beds in the shallow creeks tributary to the James River and the Bay. Although the number and extent are unknown to us, these closings appear to have been relatively small beds, though highly valuable. Official reaction from the Board of Fisheries was to sneer at those who were agitating for closure of the beds: "We cannot command language strong enough to denounce the action of some 'pure food' faddists: The scare of 'polluted oysters' has cost the workers in Virginia 'not less than one million dollars a year for 3 or 4 years'." So said the board in the annual report of 1910.[24] By 1914, it was stating that the waters of Virginia were almost entirely clean, but that because of adverse publicity, the industry was being hurt. In Virginia, as in Maryland, the concern appeared in the early days to be almost entirely about public opinion.

The board pointed out that it was difficult for a Bay native to realize that "inlanders" were indeed worried about oyster purity because of the "scare journalism" they were exposed to.[25]

The next year, the commission's tone changed appreciably. Although still labeling it "The Pollution Scare" in its annual report, it was feeling the direct effect of the marketplace, and it suggested action:

> While the polluted area is small . . . so great is its value, and so damaging its existence to the entire industry because of adverse publicity, that we deem it advisable to recommend to the General Assembly the consideration of legislation which will result in the removal of the cause of the pollution. This can be done by the towns installing sewage disposal plants.[26]

This rather indirect statement of inclination, if not intent, at least shifted the problem from the "faddists" to the pollution. No doubt the commissioners were aware of the sizable investment made by Baltimore to treat its wastes. They were at least willing to suggest that it be considered as well in Virginia.

The commission also quoted, with approval, several more emphatic statements about the need to address the source of pollution in tidal waters. Dr. Hugh S. Cumming, a surgeon with the newly formed United States Public Health Service, whose extensive surveys on the Bay area are discussed below, wrote to the chairman of the commission: "It is now recognized that no community has the right to dispose of sewage in such a way as to throw unreasonable burdens upon other communities."[27] Although this seems at first inspection to be a simple restatement of a long-established common-law principle, it was cited by the commission as a symptom that times were changing. After giving passing mention to the fact that Maryland had, in 1914, given its board of health authority to require municipalities to install sewage treatment plants, the commission quoted from a paper submitted by a Connecticut official before the National Association of Shellfish Commissioners in 1914:

> The contamination of tidal waters . . . is an evil that ought to be stopped. . . . It is a crime against nature, and it is the more indefensible because it is committed, not by savages or . . . or . . . [nationalities deleted], but by a highly cultured people.[28]

Clearly the commissioners did not wish to speak with such rhetorical passion themselves, but they were trying to make a point.

At the same time, the private oyster interests, in the persons of S. J. Watson and Frank W. Darling, sought legal remedies. Watson held a private oyster lease for some eleven acres of bottom in Hampton Creek, within the city limits of Hampton. In 1909, he was informed by county health officers that the waters were "too polluted to permit the sale of oysters therefrom," and in 1914 the state authorities, following recommendations made by a survey of federal health officials, expressly banned the sale of oysters from the creek. (The ban affected other planters in the area, and it appears that Watson brought suit on behalf of all of them.) Watson thereupon sued the city of Hampton, claiming that its discharge of sewage was unlawful and constituted a trespass against his property, for which he should be compensated. The circuit court ruled in his favor, and awarded him $4,500 in damages.[29]

The case was quickly appealed, and the Supreme Court of Appeals of Virginia reversed the decision on June 8, 1916, in a decision containing a number of important statements and principles. The decision hinged on whether the city, in dumping its wastes in tidewater, could be constrained by the private interests held by Watson. The court held that the city was exercising a public function in accordance with state law, and that this function was superior to any rights granted by the state for private oyster culture. It pointed out that the waters and bottoms of the creek, being navigable and tidal, were "owned and controlled by the state for the use and benefit of all the public." It was, therefore, for the state,

> through the legislative branch of government, to say how much pollution it will permit to be emptied into and upon waters, so long as the owners of the land between low-water and high-water mark are not injured. . . .[30]

Since the legislature had expressly authorized cities to build sewers for the disposal of wastes, and since it had not required sewage treatment or otherwise restricted the disposal of municipal wastes, the city of Hampton was within its rights and was, therefore, not liable for the damages suffered by Watson.

Here is part of the ruling:

> Since the state holds its tidal waters and the beds thereof for the benefit of all the public, we are of the opinion that the city of Hampton has the right to use the waters of Hampton Creek for the purpose of carrying off its refuse and sewage to the sea, so

long as such use does not constitute a public nuisance. . . . The sea is the natural outlet for all the impurities flowing from the land, and the public health demands that our large and rapidly growing seacoast cities should not be obstructed in their use of this outlet, except in the public interest. One great natural office of the sea and of all running waters is to carry off and dissipate, by their perpetual motion and currents, the impurities and offscourings of the land.

The state guards the health of its people for the benefit and protection of the public at large and under present sanitary standards sewerage systems for all thickly settled communities have become an imperative necessity, a public right, which is superior to the leasing by the state of a few acres of oyster land. . . .[31]

In short, the court hammered home the principle that sewage disposal was in the public interest, subject only to standards imposed by the legislature, which governed the use of the tidal waters in the public interest.

Undaunted by the Watson decision, Darling brought suit against Newport News for the pollution of Hampton Roads, where he held large and valuable oyster leases.[32] This time the circuit court decided for the city, based on *Hampton v. Watson,* and the Virginia Court of Appeals agreed. The court advanced the strong pragmatic argument that pollution was an inevitable consequence of commerce, industry, and large settlements. Therefore, in providing for private oyster culture, the legislature did so on the assumption that the right of riparians to discharge their wastes to the sea was a superior right. The court said:

The right claimed by the city clearly existed before the enactment of the oyster law cannot be doubted, and the Legislature cannot be presumed to have intended to destroy this ancient and undoubted public right in the absence of a clear and explicit statute indicating such a purpose. We think the more reasonable view of the statute, that established the private oyster program, is that it was not conceived that it would be thought desirable to continue to plant oysters in an area so certain ultimately to be polluted. . . .[33]

In other words, use of waters of the area for oystering is ". . . subject to the ancient right of the riparian owners to drain the harmful refuse of the land into the sea, *which is the sewer provided therefore by nature. . . .*"[34] [emphasis added]

Perhaps encouraged by a long and complex dissent written by one of the justices of the Virginia court, Darling proceeded to the United States Supreme Court. There, in an opinion handed down on April 28, 1919,[35] Justice Oliver Wendell Holmes made short work of the plaintiff's case, and in sustaining the decision of the state court, added the observation that reflects the prevailing attitudes of the times, as well as the law:

> The ocean hitherto has been treated as open to the discharge of sewage from the cities upon its shores. Whatever science may accomplish in the future we are not aware that it yet has discovered any generally accepted way of avoiding the practical necessity of so using the great natural purifying basin.[36]

Ignoring the experience of Baltimore or of numerous European cities in the treatment of sewage, Justice Holmes continued:

> But we agree with the court below that when land is let under the water of Hampton Roads, even though let for oyster beds, the lessee must be held to take the risk of pollution of the water. It cannot be supposed that for a dollar an acre, the rent mentioned in the Code, or whatever other sum the plaintiff paid, he acquired a property superior to that risk, or that by the mere making of the lease the State contracted, if it could, against using its legislative power to sanction one of the *very most important public uses of water* already partly polluted, and in the vicinity of half a dozen cities and towns to which that water obviously furnished the natural place of discharge.[37]

These cases, beyond laying the claims of the private oyster lessees to rest, established two things. First, waste disposal was considered to be an important beneficial use of water, a concept that may seem jarring today, but is explicitly stated in much Bay literature and popular writings at least up through the 1950s. Second, the courts made it clear that it was within the power of the Virginia legislature to require a reduction in pollution, though it had not done so.

Litigation out of the way, the situation in Virginia stabilized, so that by 1923 the commissioners of fisheries could say, "As yet Virginia has no serious question of other pollution on its oyster beds, due to the vastness of our waters and no cities or large population within our confines."[38] The "oyster scare" had become a reality for the private planters whose beds were affected, but Virginia had neither seen fit to legislate to restrict waste discharge, or to invest in municipal treatment plants.

Earlier in the twentieth century the federal government had become involved with the Bay fishery. Along with the formation of the Public Health Service (1912), the Congress authorized the conduct of studies on the pollution of navigable waters. This move led to an extensive survey of the Bay by Surgeon Cumming, who described the sanitary condition of the Bay in a report remarkable for its scope and detail.[39] The report confirmed the commonly held notion that bacterial contamination was largely confined to the waters immediately in the vicinity of populated areas, with the most confined and slowly circulating waters presenting the greatest risk of contamination of shellfish. Cumming, working with health officials in both states, identified a number of specific situations that posed a threat to human health. Many of these dealt with the actual handling of oysters once harvested, and with the practice of "drinking" oysters, which involved placing them in fresh or near-fresh water to plump them for the market (a practice which sometimes exposed the oysters to polluted water). The federal presence was, therefore, primarily of a technical assistance and advisory nature, although the Pure Food and Drug Act of 1906 and subsequent acts gave the federal health officials authority to intervene in cases that involved the interstate shipment of contaminated foodstuffs.

The Public Health Service also had to perform a sanitary survey of the Potomac estuary in 1913-14. The oyster interests on the Potomac—in both Virginia and Maryland—were concerned that the sewage from the growing population of Washington, which in 1913 numbered approximately 320,000, would destroy the purity of Potomac oysters. The survey concluded that there was no immediate risk to the oysters, since the upper limit of the beds was far removed from the Washington sewer outfall, and the net downstream transport of polluted water was slow enough to allow adequate time for dilution and oxidation of the wastes.[40] Thus, the Potomac estuary was spared the problem that was faced in the Baltimore and Hampton Roads areas in the early part of the century.

The stability of the oyster sanitation issue, which is expressed by the 1923 statement of the Virginia Commission of Fisheries quoted above, was shattered by a major outbreak of typhoid in Chicago in November and December of 1924, as well as lesser, but significant, outbreaks in Washington and New York. In all, about fifteen hundred cases were reported, resulting in 150 deaths. Since most of these cases were traced to contaminated oysters, the state of Illinois imposed an

As part of the response to public concern about disease associated with Bay oysters, federal authorities inspected incoming takes. In this photograph of 1914, an agent is at work on Long Wharf in the Inner Harbor, Baltimore. Courtesy Enoch Pratt Free Library.

immediate ban on the importing of raw oysters. The resulting publicity brought the Bay oyster industry virtually to a standstill.[41]

The Bay states, as well as the other oyster-producing states in the nation, quickly joined with the Public Health Service to establish a program of oyster sanitation that would restore confidence in the purity of oysters. The approach involved the formal adoption of bacterial standards for oyster-growing waters, as well as standards and practices for the handling and processing of oysters from harvest to the marketplace. Today this system still relies on the Federal Public Health Service to certify as to the adequacy of state programs.

Each state responded quickly to the problem. Maryland immediately moved to close beds in polluted areas, upgrade the sanitation practices of harvesters and packers, and convince the Illinois authorities that the Maryland oyster was indeed safe. To that end, the Director of Public Health for Illinois Dr. I. D. Rawlings was brought to Maryland. There he personally conducted an extensive survey of Maryland oyster waters and processing plants. He found that, with a few exceptions, the waters of the state were clean. He suggested that a set of specific practices should be adopted by harvesters, shuckers, and packers, to insure that oysters would not be contaminated. By October of 1925, the commissioner of conservation was able to report to the governor that the state of Maryland could certify its oyster safe to the state of Illinois, and "that confidence was returning to the industry."[42]

At the same time, beds in suspect or contaminated areas near Cambridge, Crisfield, Salisbury, and Annapolis were closed, and a campaign was launched to eliminate or improve toilets, privies, and pipes that were discharging sewage to tidewater. After a couple of years of effort, the total restricted area was reduced from several thousand acres to 1,288 acres by January 1, 1928, more than half of which were in the Severn River, near Annapolis.

Stimulated at least in part by the oyster sanitation issue, Maryland also launched a sizable campaign to build sewage treatment plants in its tidewater towns. By 1934, it asserted that in the whole state sixty-eight percent of the population was served by sewerage, and that the sewage of fifty-seven percent received some form of treatment, "a record not surpassed by that of any other state in the United States."[43] In that year alone, $4.4 million was spent on sanitary projects, most funded by aid from the United States Emergency Public Works and Civil Works Administrations, two depression-era sources of assistance.[44]

Virginia also moved quickly in response to the Illinois incident, but its problem and its solutions were markedly different from those of Maryland. Because the center of the Virginia oyster industry was so close to its largest tidewater metropolitan area, the stricter sanitation standards adopted in 1925 resulted in the closing of a much larger area than in Maryland, where, it may be remembered, the largest concentration of people in Baltimore were many miles removed from productive oyster grounds, and the sewage of that city had been undergoing treatment for over a decade. Virginia closed 8.3 thousand acres in 1925, 16 thousand more in 1926, and 13.7 thousand more in 1927. By 1933, a total of 38 thousand acres were "condemned," of which approximately 12,000 were considered to be productive.

Although Virginia launched a campaign to eliminate problems in nonsewered areas, it was clear that its primary problem was the result of untreated sewage discharged by the municipalities, military installations, and other government facilities of the Norfolk-Hampton-Newport News area. Thus, the problem was solvable only through the installation of sewage treatment plants. Such a solution had to gather additional political support and economic feasibility before it was finally undertaken after World War II. Although little concrete action resulted in Virginia, there was a great deal of talk, much of it of considerable historical interest.

As might be expected from previous quotations, the Fisheries Commission confined itself to lamenting the state of things that caused the closing of ". . . the greatest oyster producing territory that ever existed or will exist in the whole world." Although there were, no doubt, unrecorded efforts to address the situation, the commission appeared resigned in 1929, when it said, "It [the Commission] never expects to see Norfolk and Portsmouth provide sewer [sic] disposal plants for the relief of the shellfish industry. These cities may do it in years to come for health conditions and sanitary improvements along their waterfronts."[45] In other words, the cities were acting (or failing to act), and the oyster industry could not muster sufficient strength in the legislature to cause them to do otherwise. The next year, when the bed closings reached thirty thousand acres, the commission's apparent passivity matched its concern: "This is a problem so great that it staggers the best thoughts and energies that can be employed. . . ."[46]

The Virginia General Assembly did, however, address the issue when it appointed a commission to study the pollution of tidal waters. This group, known as the Spatley Commission, reported to the General Assembly in 1928. No direct statement of its findings or recommendations was located during this study. Whatever its findings, however, the General Assembly took no action, except that in 1930 it authorized the city of Newport News to issue bonds to construct a sewer and, if the city so chose, a sewage disposal plant, to prevent the pollution of Salter's Creek, the body of water that received most of the wastes from Newport News when Darling sued the city in 1918. The city chose not to build a treatment plant, but proceeded with plans to build a new sewer outfall in Hampton Roads some two thousand feet from the low water mark, to replace the main city discharge into Salter's Creek proper. In effect, the city was adopting the dilution strategy: it proposed to move its outfall into a body having a large volume of water and substantial currents to disperse the wastes.[47]

At this point, the executive branch in the person of the attorney general, acting on behalf of the governor, sued the city, requesting the court to require the city to install a modern sewage treatment plant, eliminate its discharge to Salter's Creek, and abandon its plan to discharge untreated sewage into Hampton Roads. That made the suit extraordinary. In effect, the executive branch was suing the General Assembly for failing to exercise its responsibility for protecting the public interest in tidal waters, specifically the rights of public fishing and the right to water clear enough for recreational use. In its brief the executive branch claimed:

> The General Assembly is the department of the government to which the administration of this property (the tidal waters and bottoms of the state) is committed, but the state as such is the trustee; and no alienation or disposition thereof by the General Assembly is legal which does not recognize and is not an execution of the trust upon which it is held for the people, nor is any use thereof by sufferance of the General Assembly legal which substantially impairs the common rights of fishery or other common rights of the people therein. It is the duty of the General Assembly to make provision for the protection and enforcement of the common rights of the people in the tidal waters. But if it authorizes, permits, or suffers an individual or municipality to use the tidal waters . . . in such a way as to destroy or substantially impair . . . the right of the people . . . it is the duty of the executive department of the state government to invoke the aid of the judicial department to restrain the individual or municipality from so doing.[48]

In a lengthy and fascinating opinion, the Virginia Supreme Court upheld the decision of the Circuit Court of Richmond, which had dismissed the suit (i.e., decided for the city). The court found that the concept of public trust, as argued by the attorney general, did not apply.[49]

The reasoning behind this opinion runs to nearly ten thousand words, and explores some dark and dusty corners of constitutional law, but in sum the court concluded: the right of fishery was something that the General Assembly could restrict or impair, so long as such action was in the public interest; the discharge of untreated wastes into tidal waters, following the arguments of Darling and Watson, was a public beneficial use of long standing; and the General Assembly, by expressly giving Newport News the authority to build a sewer to discharge into Hampton Roads, while neither expressly nor implicitly requiring treatment, "must be construed as authorizing Newport News to discharge its sewage into the Roads untreated." In short, the benefits of waste disposal and fishery are equal in status. And the restrictions and limitations to be placed on these several uses are questions committed by the United States Constitution to the discretion of the legislature free from control or interference of either the executive or judicial department of the government.[50] Clearly the fisheries interests, both public and private, would have to fight and win their battle in the General Assembly. With closures of oyster beds to reach sixty thousand acres by 1934, they must have wished for some of the political clout of their fellow watermen in Maryland, where the oyster interests had forced Baltimore to treat its sewage on the mere suggestion that it might hurt the Maryland industry.

In 1934, the legislature once again received a report from a pollution study commission, appointed by the governor at the request of the General Assembly "to study . . . the most practical and economic methods of controlling the pollution of [Virginia] waters." The commission submitted a brief and pointed report urging the creation of a "Commission of purification of waters," to be given powers to prohibit pollution, float bonds for the construction of sewage treatment facilities, and approve discharges.[51] The commission was emphatic in its position: "No language used in this report in describing the existing conditions need to be considered as too strong. It is, therefore, urged and *insisted* that the most careful consideration be given to the conditions outlined herein."[52] In outlining the problem, the commission addressed not only the threat to the fisheries industry, but warned that bathing beaches would be forced to close, "property values and tourism destroyed, health threatened, and terrible economic waste created."[53]

The passion of the majority of the commission was more than matched by the unanswered logic of a dissenting member. Mr. J. C. Biggins wrote an opinion as long as the majority's and apparently to more effect because the legislature again took no action. His argument could serve as a model for anyone wishing to oppose public investment in waste treatment, because it contained just about every imaginable argument. He first asserted that the public cost would be very high and that there was no guarantee that the situation would be much improved, in that there were numerous sources of pollution for which there was no feasible control. He cited as an example the situation in Baltimore, where, despite investments of over $20 million, one could still not swim or fish in Baltimore Harbor. Furthermore, even if substantial improvements could be made, there was a question of whether they would be in the economic interests of the area. Expensive requirements for waste treatment might put the port at a competitive disadvantage relative to more liberal (i.e., permissive) areas. Would it not make more sense, he argued, to recognize a de facto situation and zone the Hampton Roads area for commerce and industry, thereby excluding any claims against these activities by other users (except, presumably, matters pertaining directly to human health or common-law nuisance)?[54]

No doubt the issues were further debated, both in the legislature and without, but of direct evidence of that discussion we found none. In 1938, the Virginia legislature created the Hampton Roads Sanitary District, giving this regional body authority to raise money, construct and operate sewage treatment facilities and sewerage systems, and to control pollution in the Hampton Roads area.[55] And the "oyster scare" still refused to go away. In their report for 1938-39 the Commissioners of Fisheries lamented that there simply was not enough demand for the oyster crop. Production was not the problem. "There are just not enough people eating seafoods," it said. Here, to be sure, pollution had had its effect. ". . . the thing that is most seriously affecting our consumption in Virginia is the much publicized pollution conditions in Hampton Roads and other Virginia sections."[56] But what was needed was a campaign to convince people to eat oysters as they had in the nineteenth century, and to that end Virginia attempted to develop a marketing exhibit in conjunction with Maryland for the 1939 World's Fair.[57]

A patent tonger from Solomons at work at the mouth of Patuxent River.
Although this photograph was taken in January 1980, it could have been
taken any time within the last forty years—the hard life of the waterman has
not changed. Photograph by Richard McLean.

This story of sewers and shellfish is a useful paradigm. It determines the importance of the oyster as the key Bay resource. It illustrates also that it is often political influence, not technical evidence, that determines the outcome of an issue. It is the first of many Bay controversies in which the technical questions were fundamentally insoluble because since it is impossible to demonstrate that a proposed activity will not have an adverse effect, the only available proof is to go ahead with the proposal and observe its effect. Those opposing the change rely primarily on the argument that an adverse effect is possible, and that it is not worth the risk to find out whether indeed such a possibility is realized. Moreover, the story illustrates that it is perhaps public opinion that is the controlling "fact" in a given issue. The ultimate concern of the oyster interests was not that oysters would be killed or that people would get sick from eating contaminated oysters. Rather it was that people would associate oysters from the Bay with sewage, make the connection to the recently established and highly publicized link between oysters and sewage-borne disease, and then stop buying Bay oysters. What was controlling was whether *people would think* that oysters from the Bay were tainted.

Finally, this story illustrates the serial nature of environmental problems. The sewer issue began with the adoption of a public water system for the city. This system is developed not only in order to supply pure drinking water, but also for the broader purpose of improving municipal sanitation and safety through the provision of water for street cleaning and fire fighting. But problems came with the increased water supply, together with the adoption of the water closet (flush toilet), more frequent bathing, and other amenities of indoor plumbing. All of which can be seen in one context as environmental advances. Contaminated water had to be disposed of. The first approach of the cities was simply to move the wastes downstream and into the middle of the Bay. This approach ran into a competing interest involving another human health issue—the pollution of oyster beds: The remedy became the problem.

# NINE

# Don't Let the Factories In

Sails are full on the blue bay,
Men are sculling a wooden wind,
Don't take the crabs away.
Don't let the factories in.
—From Gilbert Byron's, "Tangier Prayer"
in *These Chesapeake Men*

The Bay and its tributaries have been used for waste disposal ever since the area was settled. Since the early nineteenth century sawmills, slaughterhouses, and canneries have used Bay water to flush away their refuse. Statutes were passed in both Bay states prohibiting the dumping of injurious substances into waterways.[1] These statutes were enforced by sheriffs, if at all.

In the twentieth century as industries grew in size and complexity, and as they tended to become more concentrated, they came to be seen as more of a problem, and thereby drew the attention of government. It was not until after World War II, however, that industrial pollution took on its modern role as a major villain.

The notion of industrial wastes is broad and inclusive. Industries range in the character of their discharges from tomato skins of a canning factory to the complex man-made chemicals of a plastics manufacturing plant. For this period we will also apply the term to the oil pollution caused by ships using the Bay. Given this range, however, three things stand out about the problems associated with industry: first, oil was seen initially as the major problem for the Bay proper; second, food processing plants were by far the most numerous and most noticeable sources of industrial problems; and, third, the problems other than oil tended to be associated with free-flowing streams or restricted tidal waters, rather than with the Bay itself, although there are some notable exceptions that will be discussed.

Unloading a Banana Steamer, Baltimore. By the turn of the twentieth century, Baltimore had become a busy port importing from Latin America and the Caribbean. In 1905, when this picture was taken, coffee, sugar, and guano were also imported. The masts in the background rise from Bay boats that brought fruits, vegetables, and oysters for canning, Baltimore then being the canning capital of the world. The debris to the right look like oyster shells that often piled up near canneries, but it probably is debris from the Fire of 1904 that razed most of the harborside. Courtesy the Library of Congress.

The image of a giant industrial complex hovering over the Bay, with huge pipes spewing forth wastes, is new; until recently, the industry was a small building employing a few people, and the wastes were shoveled or dumped by hand. In most cases, the wastes were not exotic and lethal; they were mostly familiar, if unsavory things—like slaughterhouse offal, brewery mash, milk by-products, or chicken scraps.

## MARYLAND

Through the nineteenth century, concern mounted for the decline of certain kinds of fish in the Bay and its tributaries. At the turn of the century, the Maryland Commissioners of Fisheries laid the blame squarely on industry, the first such public utterance that we have found:

> We also desire to express our opinion in this report as to the enforcement of local and state laws prohibiting the pollution of streams with chemicals, refuse from canning houses, sawdust, and other stuff injurious to the maintenance of good fishing in our rivers, either by anglers or net fishermen. Complaints have been pouring in upon us for a year or more that the refuse from tomato canning establishments dumped into several rivers were ruining the fishing, and the falling off in the catch dates with the location of the canneries on the river banks.[2]

Since this was a period in which truck farming and canning were at their peak in Maryland, the problems were both relatively new and intense.

The commissioners also commented on the restricted or reduced runs of shad (a topic of concern in both Bay states since about 1830) and identified pollution as one of the probable causes. Surgeon Cumming, in his 1914-1916 survey of the Bay, also made note of canning wastes, and quoted an early state health department report that described the foulness of the Cambridge harbor during canning season. "Such a situation is indefensible and should be tolerated no longer by the citizens of the city," it said.[3]

Despite this expression of concern, nothing tangible was done by the state until 1914, when the health department was given authority to control wastes from industries when their discharges constituted a threat to human health or were a nuisance.[4] This somewhat restricted authority was followed in 1917 by a statute that gave the conservation commission the authority over discharges that were injurious to fish and other aquatic life.[5] This statute, which appears to have been sometimes vigorously implemented, and sometimes virtually ignored, gave the state, in the words of one official, "powers for undertaking the great work of making the waters of the State fertile for the growth of fish and shellfish life."[6]

The conservation and health departments jointly addressed the industrial problems in the 1920s and 1930s and by the late 1930s were joined by another important participant, Chesapeake Biological Laboratory at Solomon's Island. The style of dealing with the problems can be characterized as one of cooperation and persuasion, reflected by a statement of the conservation department made in 1922. After outlining the nature and scope of the problem, the report said:

> The Commission does not in any way desire to put any restraint or additional hardship by way of expenditures upon the local capital invested [in industrial plants], but it is necessary that the rivers and streams which empty into the Bay should be cleaned of pollution . . . where pollution is found, there will be an endeavor on the part of the Commission to solicit the cooperation of the indsutry concerned to stop the pollution. By this method we believe pollution can be curbed to a large extent.[7]

The same annual report discussed the state of industrial waste management after the National Association of Fisheries Commissioners had studied the problem for years; they recommended an emphasis on reclamation or recycling. The health department also pursued this approach with considerable vigor, as reflected in its annual reports of the same period. A *Baltimore Sun* feature article written by Pulitzer Prize winner Mark S. Watson in 1930 described the progress being made in Curtis Bay, a largely industrial area south of Baltimore. It described a number of specific cases when the state health department had worked with industries and had suggested ways in which changing processes would both reduce pollution and save money.

> An important part of the work of the State Conservation Department is, of course, the tabulation of statistics about replacement and consumption of the Bay's products. Through Commissioner Swepson Earl's figures we learn that, heavy as have been the inroads on the shad, that fish is far from extinction— for 323,000 were caught in the Bay last year. And with 2,000,000 bushels of oysters produced in Maryland waters in the same

period, there would seem to be hope for the oyster industry. And with 209,000 barrels of hard crabs caught in Maryland (plus half as much more from Virginia waters which the Crisfield packers send to market) it would seem that anyone with a passion for crabs should have been able to gratify it, for crabs run 250 to the barrel, and 200,000 barrels means 50,000,000 crabs. The restaurants from New York to San Francisco have difficulty in getting New England lobsters to satisfy the demand, are using crabs in increasing quantities, so that in fact the Bay crab is gradually replacing the lobster as King of Crustacea.

In such impressive facts as these there is lively demonstration of Maryland's fight against industrialism, for that is what the whole conservation movement amounts to—whether it be protection and restoration of forests, or game, or fish, or anything else. Now, while conservation seeks to aid natural resources, industries often find that it may interfere with their profit making, temporarily at least, and so many of them, through suspicion or ignorance, or plain hostility to any new idea, try to crush it. Hence conservationists use their wits and seek to convince the industries that they will be better off, rather than worse off, through an intelligent conservation program. Thus the natural foes of conservation often are converted into advocates and allies and success comes to the campaigns to save the woodland from demolition and the water courses from ruinous pollution and their denizens from extinction. Here in Maryland these suasion methods have been signally successful, and the proof of their soundness is at hand.[8]

Curtis Bay continued to be a focal point of Maryland attention in the 1930s. Frequent fish kills and complaints from area residents kept government officials and industries searching for causes and solutions. One of the more intractable problems was a large paint pigments firm, which produced a highly acidic discharge. The health department worked with the company to develop a new process by which the wastes could be reduced. The planned changeover required, as a temporary measure, the discharge of the wastes directly to the Bay. The health department agreed wtth the opinion on the consulting engineers working for the firm, that the discharge would have no adverse effect on oysters and other aquatic life. The conservation department however, was not persuaded, and "declined permission" for the project.[9]

This situation, which was eventually resolved to everyone's satisfaction by the adoption of a new process, reflected two most significant aspects of the Maryland approach to industrial wastes during this period. First, it is clear from this episode, as well as others like it, that state officials had a powerful influence on industrial activities. While statutory authority was limited and the staff small, state officials were directly involved in approving or modifying the industrial discharges. This involvement was a far more activist program than appears to have been appreciated by later students of the Bay. It seems that the primary limitations on the program were the limited scientific and technical capabilities then available to study specific problems, rather than any fundamental limit in statutory authority or official concern.

Second, there runs throughout the interplay between the health department and the conservation department two somewhat contrasting approaches to problems. The health department took what for simplicity might be known as an engineering view, while the conservation department took a biological view of the issues at hand. The former tended to be more pragmatic and empirical, stressing doing what was feasible and then seeing what the results were, while the latter tended to be conservative, stressing *potential* damage due to little understood or unknown chemical-biological relations.

Industrial wastes from food processing plants proved to be generally less amenable to correction by the techniques developed by the Maryland official team. In many cases, there simply were not effective and economical techniques available for dealing with a particular waste. In others, the small size of the operation, and its marginal economic status, and, in the case of canneries, its seasonal nature made it difficult to require installation of an effective waste disposal system. As a consequence, distilleries, canneries, and milk processing plants were each identified, at different times, as the major water pollution problem facing the state.

By the late 1930s, however, there was a general note of optimism and progress among Maryland resource officials. With the revival of industrial activity following economic improvement in the mid-1930s, the conservation department found that it had more complaints to investigate. Far from being concerned, the department used this fact as an occasion to report:

> The Department has been very diligent, and in investigating these complaints, have notified the owners of these industries that the pollution must cease. The Department is glad to note that in every instance, the officers of the corporations readily

cooperated with the Department, and in some instances, have gone to considerable expense to see that no refuse from their factories would, in any way, pollute or destroy the fish life in the waters of the State. The Department feels that they have been successful in protecting the State's waters.[10]

A few years later the long time State Game Commissioner Lee LeCompte spoke of the great progress that had been made in eliminating industrial pollution, calling it ". . . the near extinction of a major nuisance."[11]

And the health department, in its reports, indicated the activities in the Curtis Bay area which had received the most concentrated attention in the state: "The officials of the industrial plants in the area and the Conservation Commission are to be commended for their splendid cooperation." And of industries: "It has been gratifying to note the recent interest, displayed by manufacturing officials generally, in the matter of stream pollution."[12]

In 1936, the statutory authority of the conservation department was used in court, apparently for the first and only time, resulting in the levying of a nine hundred dollar fine against the Owings Mills Distillery, Inc., for the release of caustic soda into Jones Falls.[13] The case was hailed as a landmark because it demonstrated the legal backing available should an industry prove recalcitrant. However, the prevailing effort as expressed throughout the period was to resolve industrial pollution problems through cooperation. Speaking generally, but clearly with the Maryland experience in mind, Dr. Abel Wolman summed up this approach as follows:

> If any specific feature has been responsible for progress [in water quality control] in many of the states, it has been the existence of one or more informed officials who have had sufficient energy and wisdom to carry the program forward by cooperative activity with industry and municipality. Only rarely in such progressive areas has it been necessary to invoke the law.[14]

By the onset of World War II, the Maryland program of control of industrial effluents and the building of municipal sewerage treatment plants was a source of pride. It was the opinion of many that it was a national leader in the field, and it may indeed have been. Certainly, there was a great deal of effective activity by a very small staff, and twenty-five years had seen the establishment of legal basis for action followed by a program that involved the health department, the conservation agencies, and the research establishment in a cooperative routine.

By 1945, the operating question was, "Given the successes of the Maryland program, was it enough?" A number of symptoms appear in the record to suggest that the answer was no. Perhaps the most poignant concern was expressed by R. V. Truitt, the long-term director of CBL (the Chesapeake Biological Laboratory). CBL had been doing pollution-related studies in the Curtis Bay and Baltimore Harbor areas since the early 1930s, and by the latter part of the decade was looked to by the conservation agencies as the major source of expert information. Consistently, Dr. Truitt and others at the laboratory had pointed out that the assimilative capacity of the waters of the area was great, and that there was no evidence that industrial wastes were having an adverse effect on economically important resources. In making this statement in 1940, Dr. Truitt stated: ". . . more extensive study is planned on the problem of possible *accumulative effect* of continued discharges of wastes upon *the delicate biological balance of Bay waters* and the relationship of these changes to conservation."[15] (emphasis added) Here is sounded, perhaps for the first time, the warning that has been so much a part of the ongoing debate about the quality of the Bay. The CBL scientists were aware of the large quantities of wastes being discharged to the Bay in the Baltimore area, and while they were also aware of the tremendous assimilative capacity of the receiving waters, they also wondered whether other things were going on that might have gone undetected.

Various groups of citizens were concerned about other more palpable issues. Through the first half of the 1940s, a number of incidents and conditions occurred or pertained that raised questions as to the adequacy of the state's program for industrial and municipal pollution control. The residents of Curtis Bay and Back River in particular were not persuaded that progress was being made. Newspaper accounts during that period speak of fish kills, nuisance conditions, and generally unacceptable conditions, and it can be inferred from the degree of official attention to these areas that the level of complaints was relatively high.[16] A 1942 fish kill in Curtis Bay resulted in the official explanation that it was part of a broader fish kill that was probably not related to any man-induced condition. In the same period, the health department pointed out that the unpleasant conditions in Back River were not caused by industrial pollution but by "profuse growth of algae due to high organic content of the

Field trials at the Carpenter estate on Bohemia River, Maryland, March 28, 1936. Love of outdoor sport around the Bay included duck hunting and the breeding of Chesapeake Bay retrievers. Courtesy Enoch Pratt Free Library.

effluent from the Back River Sewage Treatment Plant." And the Baltimore Inner Harbor had once again achieved notoriety as a seriously polluted body of water.[17]

More generally, indeed statewide, the conservation and sportsman's groups were starting to aggregate for more effective action against industries. In 1935, forty-four conservation organizations and other groups united to form the Maryland Outdoor Life Federation.[18] The federation promptly began to lobby for a change in the organizational structure of the conservation agencies of the state, and also for the establishment of a board of pollution control. In 1938, a committee on Bay pollution was formed, which pushed for greater recognition of potential and real problems faced by the Bay due to the growth of population and industry. The Izaak Walton League of mostly fishermen kept pressure on for the protection of freshwater streams; they were quick to use fish kills in the Bay or tangible instances of industrial pollution as evidence that a greater level of control was needed. Although the war years tended to reduce the intensity of this pressure, it was there ready to reassert itself thereafter.

## VIRGINIA

The Virginia record concerning industrial pollution during the first half of the twentieth century is not as extensive or clearcut as that of Maryland. Virginia did not have the industrial concentrations similar to those in Curtis Bay or Baltimore Harbor. Perhaps partly for this reason, it did not establish a clear legislative framework for action (as did the Maryland statutes of 1914 and 1917). Virginia's concerns during that time were apparently more directed to industrial pollution of inland streams. Nevertheless, a number of specific developments in Virginia during this period concerned industrial pollution, and many of them are important as precursors of the major Bay pollution issues of the 1960s and 1970s. Virginia's experience is, therefore, no less interesting than Maryland's, even if somewhat briefer.

Virginia officials had long been concerned over the decline of shad and other anadromous fish in her rivers. These declines were attributed to various factors, including dams, overfishing, and agricultural practices, but there was also recognition that industrial wastes were blocking the passage of fish. Numerous laws were enacted during the nineteenth century to prohibit the discharge of various polluting substances into the streams of the states. For the most part, the language specified free-flowing streams, although it is clear that one of the major concerns was the effects of wastes on anadromous fish from the Bay (and ocean) as well as concern for resident inland fish. These statutes were enforceable primarily by the local law enforcement officials; no bureaucracy was established to implement or oversee them. Characteristically, they showed more concern for the deliberate poisoning or killing of fish, than for the unintentional effects of industrial wastes, although both categories were covered.[19]

In the first part of the century, the attention that was drawn to the wastes from a pulp mill in Richmond resulted in an extensive study of the situation that recommended a combination of civil common-law remedies and new legislation to alleviate what were considered to be unacceptable nuisance conditions.[20] The concerns that prompted this study were primarily those of residents along the banks of the river, rather than those of concerns for the fish life in the river. Nonetheless, this study, which resulted in no definite action, represented the first investigation of tidewater industrial pollution located during the course of the research in this book.

With the marked increase in shipping in the lower Bay, the problem of oil from ships came into prominence and for a number of years was seen as the most serious pollution problem in the Bay. The oil came from the practice of ships pumping ballast water from their bilges as they entered the Bay to take on cargo. The problem in the lower Bay was aggravated by the presence of a large number of Navy ships. Also, ships heading for Baltimore would begin to pump their bilges as soon as they entered the protected waters of the Bay from the open ocean.[21]

The concern was felt by resort and property owners along the shores of the lower Bay, as the many fine beaches of that area were occasionally fouled by oil. But it was the Commissioners of Fisheries who expressed the strongest opposition to this practice. Although their attention had been directed to the oyster sanitation issue for some years, in 1919 they were to say:

> One of the greatest questions for the future is that of pollution. The pollution by sewage of our ever increasing population and the waste from our rapidly growing industries is affecting the entire fish and oyster industry in and around Hampton Roads, *but of all the destruction caused by pollution that from oil waste is the worst.* (Emphasis added.)[22]

They pointed out that oil, unlike sewage, affects not only the mature oysters, smothering them or making them unpalatable, but it also interferes with the plankton that is the food for oysters, and directly kills oyster spat when it is part of the plankton (plankton are floating plant and animal organisms that are largely incapable of motion; oyster spat larvae are part of this plankton until they settle and attach to some hard surface on the bottom).

Thus the commission pinpointed oil as a clear enemy for the organisms of the Bay. This concern was shared by the other eastern states, including Maryland, and their combined action led to the passage of the Federal Oil Pollution Act of 1924.[23]

How effective this act was is not discernible from the records we examined. Its passage, though, coincided with the "Illinois incident," which quickly vaulted oyster sanitation ahead of oil as a major concern. The fisheries commission, after commenting on it for a number of years, gave it no mention in its annual reports thereafter. It did receive mention in a 1935 report by the Virginia Planning Board as a significant problem, and it was still a major issue in Maryland following World War II. For this period, however, it was an issue that both states conceded was the responsibility of the federal government. It, therefore, marks the first instance of a direct regulatory role by the federal government in a water quality issue of the Bay. This modest beginning, of course, gives no clue as to the major growth of federal involvement that took place in the later part of the 1960s.

In 1930, a massive die-off of oysters in the York River began a controversy about the effects of industrial wastes that persists to this day. A large pulp mill at West Point was suspected of being the cause of the die-off.[24] Government officials in Virginia and their representatives in Washington sought the aid of the United States Bureau of Fisheries, which established a laboratory at Yorktown "solely for the investigation of industrial pollution in the York River."[25] The study was carried on for most of the next decade. Although oyster productivity declined in the area, and the pulp mill wastes were "definitely implicated," there was no government action taken to abate the wastes while the scientists sought to determine the specific causal agents of the decline. Government at the state and federal levels apparently felt it necessary to establish a direct link between the discharge and the observed declines, before requiring remedial action. Such proof was not forthcoming.

Deerfoot Cook, chief of the Pamunkey Indians, shows off his catch of shad. Shad and other fish were the major source of income of this Virginia tribe as late as 1960. Courtesy A. Aubrey Bodine Collection, the Peale Museum.

What did emerge from the study was the recognition that the state needed a research capability to address the kinds of questions raised by the York Pulp Mill, as well as to address the broader questions of management of Bay species. After the federal government pulled out of its York River studies, the state of Virginia and the College of William and Mary established the Virginia Fisheries Laboratory at Yorktown, with Curtis L. Newcombe as its director. The laboratory took up from the Marine Biological Laboratory the task the latter had adopted in its last years, which was ". . . to find ways and means for improving the tidewater fisheries resources of Virginia. . . ."[26] Thus, while the oyster die-off on the York remained perplexing, the trade wastes of the pulp mill led chronologically, if somewhat tangentially, to the state laboratory, later renamed the Virginia Institute of Marine Science, that was to play a major role in pollution issues after the war.

Virginia also faced an industrial issue in the late 1930s and early 1940s, the resolution of which freed her from difficulties that were later experienced by Maryland. The issue was dredged spoil disposal, a perennial matter of concern on the Bay. The approach channels of Hampton Roads and the port areas required periodic dredging, both to remove accumulated sediment and to increase the size and depth. Prior to the 1930s the practice in the lower Bay had been to dispose of the dredged material overboard in the shallow areas of the Bay, where it would be distributed by the currents and eventually settle out. Watermen and property owners began to object to this, claiming that it caused adverse effects either through mechanical smothering or it transferred polluted bottom sediments from the harbor area to the open Bay, where it would contaminate oysters and four beaches.[27]

In 1940 the Congress directed the United States Army Corps of Engineers to study the problem and develop a solution. The Corps came back in 1944 with a proposal to dike a large area on the south side of the James River just to the west of the Elizabeth River opposite Norfolk. The diked area would receive the dredged spoil from all the public and private navigation projects in the area. At a public hearing on the project in 1944, some nearby oyster lessees objected that the effluent from the spillways draining the project would pollute their beds. On the basis of this objection the Corps required that the state cancel the nearby oyster leases and hold the federal government harmless for any damages done to oyster grounds in the vicinity. The project was approved by Congress in 1946, and built between 1954 and 1957.[28]

By so resolving the issue of dredged spoil disposal Virginia interests were able to eliminate it as an important issue thereafter, thus avoiding one of the most vexing and time-consuming controversies that faced state and federal officials in Maryland from the early 1960s to the present.

The experiences of the two states with industrial pollutants shifted the perspective of officials and scientists with respect to the Bay. Pollution had hitherto been seen as an engineering problem. The questions had been: "How to keep the shipping lanes free?" or "Where to place sewage outfalls?" But as ever-increasing industrial discharges killed fish, both states established fishery laboratories. Scientists at these laboratories looked at the Bay as a biological system.

As early as 1933 a consensus had developed that the Bay must be considered as a single resource unit. In October of that year, an interstate conference on the Bay was held,[29] the first of many conferences to discuss the management of the Bay. The conference proceedings are of interest in that they give a reasonably accurate picture of what was on the minds of the resource managers of the time. Most strikingly, a number of speakers sounded a theme so common today—that the Bay is a national treasure, that it must be viewed and managed as a single system, that man's activities must be carefully monitored and controlled, and that some sort of interstate body should be established to deal with the management and protection of the Bay. Although by no means all speakers addressed these issues, they were stated often enough to suggest that such ideas were by no means rare or radical.

By the early 1940s the perspective of the Bay as an ecological system was well established. Canadian scientist A. G. Huntsman, who was working at Maryland's Chesapeake Biological Laboratory during this period, expressed it well.[30] He suggested that the Bay was an extremely complex and open system, and that long and careful study would be necessary before it could be managed intelligently as a unit. In the meantime, however, it would be necessary to have ". . . investigators prepared to brave the criticisms of the academic theorists. . . . These investigators should not hesitate to draw preliminary conclusions from limited facts. . . ."[31] In other words, although science was going to progress slowly in unraveling the secrets of the Bay, the needs of management for scientific guidance would not wait, and, therefore, there needed to be courageous "directed research."

Huntsman also stressed the openness of the Bay system. He discussed the influence of land drainage from the large watershed area draining to the Bay, and pointed out that much of what went into the Bay started its journey from well outside the tidewater area. He also reminded his readers of the ocean connection, saying: "The Bay is very far from being a discrete productive unit, since there is said to be extensive movement of the fishes out from and back into the Bay."[32] Of man's activities, Huntsman sounded a rather striking view, one related directly to the controversy then underway with regard to the effects of the York pulp mill effluent:

> It is perhaps axiomatic that the changing character of the fisheries of the Bay is determined mainly by changes in the physical chemical conditions of the Bay. But we are prone to think first of man's actions, of the psychology [physiology?] of the fish, and of biological factors, and to overrate the probability of these being responsible for any changes in a fishery.[33]

By 1945 significant advances had been made in efforts to control industrial wastes. Particularly in Maryland, government officials convinced industry to make in-plant changes to reduce discharges. And there had been numerous attempts to address the difficult scientific, technical, and economic problems posed by industrial effluents and their control. But despite this activity, a clear opinion evolved that the situation was growing steadily worse. Growth in population and industrial activity had simply outstripped the activities of the governments of the two states. Whether there were actually less or more wastes reaching the Bay in 1945 than in 1900, our research would not allow us to say. Whatever the facts, there was definitely a growing concern among the citizenry of both states, particularly among sportsmen's groups. Pollution of the Bay, and more specifically of many of its tidal and nontidal tributaries, was seen as a major and growing problem.

# TEN

# A Bay Bureaucracy

Rapid growth of population and industry is outstepping
the ability of the State Water Control Board . . . to deal with
the problem.
                              —Commissioners of Fisheries of Virginia,
                                  1959

World War II diverted the attention of the body politic away from
Chesapeake Bay. In its aftermath, however, both states created new
administrative agencies and charged them with responsibility of
controlling pollution on a sustained basis.

When the war ended in 1945, Virginia had in place the Hampton
Roads Sanitary Commission, and this commission immediately resumed its work to sewer and to treat municipal industrial wastes in
this prime Virginia trouble spot on the Bay. By 1947 the commission
had committed $12 million to the task, and other improvements to
municipal systems, actions by industries, and the correction of some
of the smaller sources of pollution in restricted waters resulted in the
reopening of oyster beds.[1] At the same time, the Virginia Fisheries
Laboratory resumed its study of the biology of the Bay, and also
turned its attention to exotic pollutants such as DDT.[2]

The most important development of the period occurred far from
the Bay in the vicinity of Front Royal, Virginia. A plastics plant had
been constructed on a branch of the Shenandoah River, and its discharges, begun in 1940, had completely killed off the aquatic life on a
sizable stretch of the river. Fishermen and conservation groups in the
area were outraged; they proceeded to seek political support for
government action. An influential state delegate from that area, A.
Blackman Moore, agreed to serve on a study committee to investigate
what needed to be done about water pollution in the state. His
committee came forward in 1945 with a recommendation that a state

agency be established with responsibility for the management of pollution control throughout the state (excluding the area under the jurisdiction of the Hampton Roads Sanitary District). The bill went before the 1945 session of the General Assembly, and under Moore's protection and influence, passed the house and senate with only a few dissenting votes. One of the features that allowed for broad accept-ance was a grandfather clause that exempted industries and munici-palities already in place. Any expansion or modification, however, required regulatory approval by the Water Control Board. The basic approach of the board was to deal with the condition of the stream or receiving body of water, not with the discharge itself. According to the prevailing dilution theory, this approach allowed for the max-imum beneficial use of the water for waste disposal as well as for other uses. The board, therefore, had to establish that there was a harm to other uses caused by waste disposal before taking direct remedial action.[3]

Both the creation of the Water Control Board and the progress of the Hampton Roads Sanitary District met with enthusiasm from the Marine Resources Commission. In its 1948-1949 report, it said, "The problem of pollution is, we think, a vanishing one. . . . We can envision the time in the not too distant future when pollution will no longer be a problem in Virginia, and valuable oyster ground formerly condemned for use will be restored to production."[4]

The next few years seemed to support that optimism. The treat-ment plants put on line in the Hampton Roads area allowed for the opening of nearly twelve thousand acres of oyster ground, a success that was hailed as a national model.[5] The Water Control Board, using a similar approach in other parts of the state, concentrated on the development of primary treatment plants for municipal systems. Industries were encouraged or pressed to adopt internal process changes to reduce their loads to tidewater and feeder streams.[6]

Through the 1950s, however, the optimism of the Marine Re-sources Commission eroded. Read in sequence, their biennial com-ments on the pollution issue sound like a record winding down. In 1953, they said: "The problem of pollution continues to be a serious one. However, the State Water Control Board is making progess in the abatement of pollution for certain streams and the prevention thereof in others."[7] By 1957, the statement was a bit stronger: "The problem of pollution continues to plague the seafood industry despite the fine work of the State Water Control Board and the Hampton Roads

Sanitary District. The discharge of industrial waste . . . poses a real problem."[8] By 1959 it was:

> Pollution is a continuing and rapidly growing serious problem which requires prompt attention. . . . Rapid growth of population and industry is outstripping the ability of the State Water Control Board and Hampton Roads Sanitary District to deal with the problem.[9]

And, by 1961, the Commission lamented:

> Contamination of our natural waters by pollutants of various types is one of the most pressing problems facing our Commonwealth today. Pollution is directly affecting our marketing and the consumption of oysters and clams. It is very possible that it may be one of the reasons for our decline in production.[10]

This time there was no mention of the otherwise fine effort of its sister agencies.

A number of factors contributed to this decline in confidence. There was a marked growth in industrial activity in the tidewater area, particularly in the Hopewell area of the James River. Moreover, the research work of the Virginia Fisheries Laboratory identified new concerns. The laboratory investigated pollution sites and fish kills in the early 1950s and reported on the difficulty of establishing a direct link between discharges, such as those from the major pulp mill wastes at Yorktown, and deterioration in the resource. Spokesmen for the laboratory pointed out the difficulty of doing research in marine environments, when most of the national research upon which standards and pollution control were based had been done on freshwater bodies of water.[11] In its 1956-57 report, it presented a laundry list of things to worry about, while stressing that there was too little knowledge upon which to base sound and enforceable control requirements. It warned of the problems of subtle effects on nursery grounds and spawning areas; of detergents and industrial chemicals that passed through municipal treatment systems unaffected; of toxic insecticides and weed killers that could have major and persistent effects in small considerations; and of new and poorly understood chemicals, the toxic effects of which had not even been guessed at, let alone studied.[12] More generally, it described the need to study and understand the Bay as a system, in order to be able to assess the effects of man's activities. "There must be more research on pollution," the laboratory urged, ". . . before the Tidewater area undergoes an industrial explosion. . . ."[13]

A University of Maryland scientist aboard research vessel *R. V. Orion*, off Calvert Cliffs in January 1982, taking a sample from the Bay bottom. He is using a hydraulically operated clamshell dredge to collect sediment, which he will then analyze for toxic metals, radioactivity, and pollutants. Photograph by Richard McLean.

One problem that drew scientific attention was a massive oyster die-off in the Rappahannock River in 1955, repeating an earlier die-off in 1949. Once again the watermen applied common sense and argued that it was caused by upstream pollution. The scientific evidence, much of it gathered and evaluated by the Chesapeake Bay Institute, pointed to naturally occurring low oxygen levels brought on by heavy rainfall. In describing the low oxygen conditions found in various parts of the Bay, the Virginia Fisheries Laboratory stated, "There is no evidence whatsoever that pollution contributes significantly to any of these situations."[14] The director of the Chesapeake Bay Institute, in commenting on the results of its studies, expressed sympathy with the watermen in finding it difficult to accept that the problem was a naturally occurring one, since they had seen other periods of heavy rains in which such a die-off had not occurred. The evidence, however, still did not point to any man-induced changes.[15]

In Maryland, support for a pollution control agency had been developing ever since the late 1930s, especially from the conservation and sportsmen's groups. In January of 1945 representatives of the Izaak Walton League and the League of Maryland Sportsmen met with the governor and key legislators, to discuss problems pertaining to pollution in free-flowing streams. The Health Department was brought into these discussions, but the conservation agencies, which had undergone some reorganization in 1939 and 1941, were not.[16] The Board of Natural Resources formed its own study committee, and joined the private groups in urging the governor to form a study committee to examine the state program and see what changes, if any, were needed.

By late 1945, a Committee on Water Pollution was formed, and a year later proposed the formation of a commission similar to that existing in Michigan. The commission would be a coordinating body which drew together the various functions and interests of the Health Department and the fisheries interests within the Board of Natural Resources.[17]

By the time of the 1947 legislative session, the proposal had changed to creation of a new agency, similar in some respects to the board already in place in Virginia. Sportsmen's groups in particular had argued that a coordinating body would not be sufficient. The need, they felt, was for an agency that would have pollution control as its sole function. Against the objections of municipal industrial interests, as well as from government officials,[18] the bill passed, and the Water Pollution Control Commission was established.[19]

The work of the commission in the early years is well documented in its annual reports. The commission was pragmatic and modest,
stressing the twin principles of cooperation and reasonableness that
had guided the work of the health and conservation departments in
the 1920s and 1930s. The annual reports and other documents as-
sumed that waters have a natural assimilative capacity, and that it
was only after that capacity had been exceeded that interference with
other uses occurs. It was the duty of the commission to see that such
excesses did not occur, and when they did, to take corrective action.
Throughout the process, economic reality had to be kept in mind, and
requirements had to be ". . . not only technically attainable but also
financially feasible, so as not to impose upon the State's industrial
and municipal economy an expense which is out of proportion to the
benefits sought."[20] The commission stressed that its work was in the
best interests of the economy. It pointed out that it was often to the
advantage of business to support pollution abatement programs: "An
active water pollution control program . . . is a good business policy
for the State as a whole"[21] moreover, "No one reading the technical
journals of industry could fail to be impressed by the sincerity of
industrial management in its efforts to solve the problem of stream
pollution."[22]

During this period the regulators spoke well of the regulations.
With a cooperative spirit, the regulators set out with a small staff to
attack water-quality problems. These problems were seen to be
wastes from canneries, wash from sand and gravel operations, and oil
discharges on the Bay. Work on the first two concerned nontidal
streams, although most of the operations were in the coastal plain
and, therefore, had at least an indirect effect on the Bay. The oil issue
was attacked primarily by enlisting the aid of Bay pilots, who were
urged to impress upon the ships' captains the importance of observ-
ing the Federal Oil Pollution Act of 1924. Although prosecution
under the act was difficult and the fines small, a real hardship could
be placed on a ship by requiring her to remain in port while an
investigation was conducted. This tactic apparently helped reduce
the incidence of violations, because oil disappeared from the list of
primary pollution concern in the Bay.[23]

If progress was the official view expressed in the annual reports of
the Water Pollution Control Commission, the Health Department,
and the Board of Natural Resources, the newspaper coverage of the
time noted little improvement. For example, Curtis Bay and Back

Culm piles, or coal refuse heaps, showing erosion at Shenandoah, Pennsylvania, 1936. Courtesy National Archives.

River continued to be sources of criticism and complaints. In 1948, over fifteen hundred residents of Back River assembled to protest water-quality conditions there, and agitated for the city of Baltimore to do something about the Back River Sewerage Treatment Plant. The city was examining a variety of alternatives to upgrade the plant when a circuit court judge ordered the city in 1949 to consider direct discharge to the Bay as an alternative. In so doing, the judge seemed to vindicate the Sewerage Commission of 1899, when it had pressed unsuccessfully for this use of the Bay.[24] Industrial discharges continued to be complained about in Curtis Bay. And in 1955 Anne Arundel County took the unusual step of suing the DuPont Company to enjoin it from building a discharge pipe in Chesapeake Bay, although the project had the approval of the Water Pollution Control Commission.[25]

In the mid-1950s numerous communities in the upper Bay were still discharging sewage raw to the Bay, prompting the formation of the Upper Chesapeake Watershed Association, a citizen's group whose primary mission was to create pressure for the construction of waste treatment plants. Also in 1955, the first major Bay pollution news article appeared in the Baltimore *Sun*, itemizing the various problem areas, the sources of pollution, and the lack of treatment implemented by either municipalities or industry. This influential article as excerpted below suggested that the Water Pollution Control Commission was something less than aggressive in having taken only four industries to court in the eight years of its existence:

> Raw sewage is a disgusting subject. Few people would prefer to talk about it if it could be avoided. But in many parts of Maryland today, raw sewage is something that cannot be avoided. People are being forced to live with it, and their objections are becoming increasingly audible. The facts are these.
>
> Twenty-five Maryland cities and towns have public sewers but no treatment plants. The daily outpourings from the hundreds upon hundreds of homes in these communities are dumped raw into the nearest streams.
>
> Another thirty-two Maryland towns, of sufficient size to have significance from a public health standpoint, have no sewers at all. The raw sewage goes into the ground, to the extent that the ground can absorb it: or it is privately piped or channeled to nearby streams, or it overflows from cesspools and stands in fetid low places.

Still another sixteen Maryland cities and towns have sewers and treatment plants, but they are inadequate to meet the demands made upon them, so that some raw sewage either gets only partial treatment or bypasses the treatment plants altogether and flows directly into streams.

To this list of seventy-three towns must be added the Baltimore metropolitan area, where sewer systems have failed to keep up with population growth, and also those Maryland suburbs of Washington which help to pollute the Potomac. And finally, there are institutions like the Perry Point Veterans Hospital, which empties the waste of 1,500 patients and staff into the Chesapeake Bay.

Raw sewage is not the whole problem, either. Into Maryland streams and Chesapeake Bay go the waste materials of big city and small-town production: acid mine waters, toxic chemicals, offal from meat and poultry packing houses, pulp and seeds from canning companies, the washings from dairies, oil, grease, coal dust, pulp fibers, clay particles, and other foreign matter, to say nothing of the trash and garbage that householders toss into rivers and brooks.

Fifty years ago the disposal of human and industrial wastes was no great problem. The streams used for waste disposal could clear themselves of contamination in short order, using nature's own remedies. Today, however, Maryland is too populated for casual waste disposal.[26]

Also during the same time (mid-1950s) the municipalities of the Bay area were complaining that the health department was being unreasonable in its demands that they install modern sewage treatment facilities. The towns simply could not afford it, they claimed.[27] These protests that not enough was being done joined those of citizens and conservation groups.

Throughout the period, new water-quality concerns were added to the long-standing issues of municipal and industrial discharges. Wastes from boats became an issue that received mention in the late 1940s, particularly in the crowded and poorly circulating subsidiaries of the upper western shore, such as the South, the Severn, and the Magothy rivers in Anne Arundel County, and the Back and Middle rivers in Baltimore County.[28] Growing awareness of pollution as an issue was matched by rapidly growing use of the Bay for recreation. Indeed, recreation users formed a tight causal cycle, as the very boaters and residential shoreline owners who were expressing concern about pollution were also seen as sources of the problem.

Another new issue was the use of the Bay for explosives testing. Although not specifically a water-quality issue, the testing of explosives created a number of highly visible fish kills, roiled water, and created a general nuisance that was complained about by citizens and caused worry among scientists and managers. The official view was that the military activity was of localized effect and did not constitute a major threat to the Bay. Nonetheless, it was so extensive and so prominent that it came to be seen in some circles as a major menace that significantly interfered with the productivity and enjoyment of the Bay.[29]

Acid mine wastes were also a new concern of the 1950s. Coal mining in the upper reaches of the Potomac and the Susquehanna had created hundreds of miles of highly acidic and biologically dead streams. Acidic waters were for the most part buffered by the time they reached the Bay. Popular suspicion, though, held that the evil-looking "yellow boy" in mountain streams would eventually work its way to the Bay.

In 1954, the United States Bureau of Mines developed a plan that would have brought mine wastes directly to the Bay. In what is probably the biggest of the "big pipe" proposals for waste disposal in or to the Bay, the bureau proposed that a 180-inch pipe be built from the anthracite coal region of northeastern Pennsylvania to the Bay, in order to drain the underground coal mines that were rapidly filling with water.[30] The water could not have been pumped into nearby streams because of its high acidity, but the bureau argued that the Bay, with its enormous quantity of water, would quickly neutralize the acids of the mine drainage, and suffer no harm.

Maryland officials responded with outrage. After assuring watermen and sportsmen's groups that the state of Maryland would not tolerate such an invasion from a neighboring state, the governor himself held meetings with the congressional delegation and with officials in the Department of the Interior to seek to get the project withdrawn.[31] The alacrity of the response was perhaps unnecessary, as it appeared that the proposal had serious technical and economic flaws. Pennsylvania officials were not themselves enthusiastic about the project, so that by the end of the year it had died. Nonetheless, it is a project of considerable interest because it reflects another example of the attractiveness of the Bay as a sink for wastes, and it represents one of the first clear examples of an interstate water quality issue.

A second "big pipe" proposal marked this era, this one with a source closer to the Bay. Consulting engineers working for the District of Columbia proposed that the municipal wastes from that city, which contained a substantial loading from the Maryland suburbs, be piped across southern Maryland and discharged into the main stem of the Bay. This would remove pressure from the highly restricted and slowly circulating Potomac River estuary, and would provide a source of irrigation water to southern Maryland farmers, then experiencing a rather severe drought. In addition, suggested the consultants, the wastes would provide additional enrichment to the Bay, which would be beneficial to the total productivity of the Bay and not injure any of the Bay's resources.[32]

Hence the decade of the 1950s was marked by two major developments. Both states created water pollution control agencies and charged them with the job of cleaning up Bay waters. These agencies sought to cooperate with industry and municipalities in developing financially "reasonable" pollution control strategies. These efforts first met with enthusiasm, but as the decade progressed first conservation groups and then the agencies themselves seemed to lose confidence in their ability to control pollution.

# ELEVEN

# Save the Bay

You and I differ principally in philosophy, and this should not be misconstrued. You are inclined to encourage full use of the Bay unless available knowledge proves that human uses will be impaired. I am more conservative and prefer not to risk damage to the Bay until reasonably good estimates can be made of all effects.

—L. Eugene Cronin in a letter to Donald W. Pritchard, March 6, 1969

Clustered around 1960, a number of events presaged fundamental changes in the kinds of issues that concerned Bay governments. Virtually all of the previous concerns were prompted by episodes when municipal and industrial wastes at various locations attracted the attention of scientists, administrators, legislators, and the courts. After 1960 concerns became larger. The concept of water quality became all-inclusive; governments more and more asked the question: "What is the condition of Chesapeake Bay?" The Bay became generally recognized as an ecological system, subject to a wide variety of man-made influences transcending the boundaries of states, the jurisdictions of agencies, and the capacities of science to define and predict.

Some issues of the 1960s had been around for some time; others were new. But all were marked by problems of increased size, complexity, and persistence. Flow regulation or modification of the major rivers (Susquehanna, Potomac, and James) as a tool in water quality management, came under consideration. The effects of waste-heat discharges, particularly from power plants, began to be seen as the major threat to the Bay. Spoil disposal in the Maryland portion of the Bay developed into a full-blown controversy. Overenrichment of the Bay and heads of the major subestuaries became a matter of concern. The effects of navigation projects on salinity, first on the James and then on the old Chesapeake and Delaware Canal, became a point of debate. And large-scale engineering projects raised the question of cumulative effects of all of man's activity. Rapid development in the Washington suburbs and its effect on sewage disposal and soil erosion introduced land use practices as a source of concern.

Tying all of these issues together were a series of journalistic accounts of the Bay and its problems. These articles, both news and feature, began in the early 1960s and reached a peak of volume and urgency around the end of this study in the early 1970s. To the detailing of these issues we now turn.

In the early 1960s, the programs of the two state water pollution control agencies continued substantially as before. In Virginia, consensus noted significant progress on the municipal wastes front, a reduction in oil pollution, and the need to work more directly with industries to reduce wastes through the construction of waste treatment facilities. Previously, industries had been urged to adopt internal process changes that would reduce their discharges.[1] In Maryland, programs continued to stress the need for cooperation with industry in setting realistic goals and providing economically feasible solutions. Toward that end, the Maryland commission began a series of annual conferences to discuss waste control technologies.[2]

Nonetheless, there were signs that the rapid population growth throughout the region was making existing programs inadequate. The Virginia Fisheries Commission, true to form, continued to express concern for the effects of pollution, while praising or at least encouraging the Water Control Board and the Hampton Roads Sanitary Commission. "Contamination of our natural waters by pollutants of various types is one of the most pressing programs facing our Commonwealth today," it said in 1961.[3] And through the decade, while varying its primary concern from industrial to municipal pollution, it was consistent in the opinion that the problem was grave. In 1971 it repeated its 1961 refrain, "contamination of our natural waters by primary and secondary pollution is one of the major problems confronting the seafood industry . . . our paramount concern at the present is the problem of municipal sewage."[4] Treatment plants generally were not able to handle storm loads, and, as a consequence, beds were frequently closed for some time following rains. And in 1972:

> We know that the Water Control Board is working diligently to prevent contamination of our waters, but this is a gigantic task. .

A man crabbing near the mouth of the Chesapeake Bay, Lynnhaven Inlet,
Virginia, July 1975. Photograph by Richard McLean.

.. The Commission prays to the General Assembly to hear the plea of an endangered industry and to act with courage and decisiveness in providing the impetus for swift correction action in this area.[5]

In Maryland, the fisheries industry was optimistic. The areas closed to shellfish were considerably smaller than in Virginia, and the MSX disease that had greatly reduced oyster stocks in Virginia had had little effect in Maryland. The 1960s also marked a time of rapid increase in Maryland oyster harvest, due to an aggressive state-sponsored reseeding program, and several years of above-average natural reproduction. Thus, Maryland was able to recapture first place in national oyster production, a spot it had relinquished to Virginia in 1930.[6]

Maryland was also successful in upgrading its municipal waste treatment plants, so that by 1969 it claimed that approximately eighty percent of the Bay area sewage received secondary treatment.[7] And on the industrial front, there was a consistent sense of progress in the reports of the Water Pollution Control Commission (renamed and reorganized as the Department of Water Resources in 1964), where the number of orders for compliance, investigations, court actions, and other measures of activity continued to increase.[8] These official reports are less than convincing, however, since they give only a numerical report on activity, and do not reflect the volumes of wastes or the actual conditions of the receiving waters.

A significant change in the federal role in water quality control came about with the passage of the Federal Water Pollution Control Act of 1964. This act required that the states establish water quality standards for all interstate waters within their borders, in order to qualify for federal assistance in waste treatment financing, and to avoid direct federal enforcement intervention. Both Maryland and Virginia developed programs of compliance and, in the process, underwent rapid growth in their water control staffs and budgets. The direct impact of this administrative and regulatory change on the Bay is difficult to discern, but it is noteworthy that the shift to water quality standards did not produce a record by which net progress could be measured (despite the emphasis on the quality of the receiving water rather than on control of discharges). The only available benchmark for achievement of conventional water quality objectives (i.e., control of discharges from industries and municipal sewer systems), appears to be in the summary comments of the numerous journalistic discussions of Bay water quality during this period, which will be discussed below.

## SALINITY MODIFICATION

In the early 1960s, two Corps of Engineers' proposals for major public works projects on the James River introduced salinity modification as a "water-quality" issue and prompted scientific study and political controversy. The first was a river basin program for the entire river, issued in 1962.[9] The multipurpose program, consisting primarily of a system of dams, would have allowed for the partial control of fresh-water flows into the estuary, thus affecting salinity. The second project proposed the deepening of the shipping channel to Richmond from twenty-five to thirty-five feet.[10] Because salinity distribution in the Bay and its tributaries is affected by the size and depth of the channels, this project also suggested that a change in salinity would occur.

Salinity had long been a matter of concern to Virginia oystermen. At the low end of the range of the oyster, an unusual rise in freshwater inflow will depress salinity and result in stress or death to oysters. This change was often experienced on the Potomac, and to a lesser extent on the other major Virginia rivers. Generally of more concern, however, were the higher salinities, because numerous pests of oysters were more prevalent in the higher salinities. Other things being equal, it was and is desirable to keep maximum salinities low, especially in the vital seed-producing portion of the James. The invasion of MSX in the late 1950s, which had hit the Virginia oyster industry so hard, underscored this relation. MSX was generally limited to higher salinity waters.[11]

In considering these two Corps projects, then, Virginia officials and watermen attempted to assess the impact of the proposals on the health of the oyster industry. This concern, of course, was nothing new. What was novel was the fact that they were now dealing with a fundamental parameter of the natural system, salinity, rather than with a relatively small quantity of a constituent added by man. The concern had enlarged from pollution to basic ecology. Obviously, significant changes in the salinity of the James could have substantial effects on the entire biological system supported by the estuary. The question was, what were desirable, and what were undesirable, changes?

The evaluation of the proposed river basin plan led to a generally favorable response from the tidewater interests. The result of the program would have been to even out the flow of the river over the course of the year. That evening out would prevent summer salinities from going too high in the lower estuary, thus holding the high salinity predators at bay. It would also reduce the heavy slugs of floodwaters, which tended to limit oyster production in the upper part of the estuary. Two related water quality benefits were expected. Increased summer flows would provide more dilution waters for the wastes entering the river at Richmond. This increase would tend to lessen the degree of pollution downstream in the freshwater portion of the estuary. Second, the reduction in floodwaters would also reduce the oxygen-depleting effects of storms, which had been held accountable for heavy oyster mortalities in the Rappahannock River during the previous decade.

The proposed shipping channel was quite another matter. Largely through the careful work of the Chesapeake Bay Institute, it was understood that the James River, as well as the Bay and the other major tributaries, had a two-layered circulation pattern: a net upstream flow of dense high salinity water in the lower part of the water column, and a net seaward flow of fresher less dense water above. Deepening and widening the channel, it was reasoned, might increase the total volume of seawater movement upstream and result in higher salinities. It would also physically aid the upstream movement of oyster predators. The fisheries interests, faced with a further threat to their depleted industry, vigorously opposed the project.[12]

After a couple of years of jockeying between the fisheries and port interests, a positive approach to resolving the conflict was adopted in 1964. The Virginia General Assembly appropriated $400,000 for the construction of a hydraulic model of the lower James estuary. This model to test the effects of the construction of the navigation channel on salinity distributions was quickly constructed at the Corps of Engineers' waterways experiment station in Vicksburg, Mississippi. Using Corps experience in calibrating and testing such a model, together with inputs from the Bay scientific community, it was determined that the channel would have virtually no effect on the prevailing salinity regimen of the estuary. On the basis of that finding, the governor was authorized by the General Assembly to give consent to the project.[13]

Hydraulic model of the lower James River estuary, built in 1964. Pictured here is the Chickahominy River reach. Courtesy College of William and Mary, Virginia Institute of Marine Science, School of Marine Science.

What had been a major standoff between competing interests was resolved by using the best technology available to make an informed judgment as to the effects of the project. The report stands as a landmark of rationality and objectivity in Bay resource management decision making. Regrettably, from the standpoint of completeness, the navigation project was never begun. During the time that the project was in doubt, an oil pipeline was built between Hampton and Richmond. Since oil was one of the commodities that gave the project its economic viability, the provision of an alternative transportation mode changed the picture. Upon reevaluation the Corps found the project to be only marginally in the black (with a benefit/cost ratio of barely over one), and so the project was withdrawn. It should be noted that there were other environmental problems associated with the project, which would have become more prominent had it been pursued. Chief among them was the problem of spoil disposal, which had long vexed Maryland officials. Since that problem came to a head in Maryland at about the same time that the James River controversy was underway, we will now turn to it.

## DISPOSAL OF SPOIL

Navigation channels and facilities in and around Baltimore Harbor had required frequent dredging since the early part of the nineteenth century. The Patapsco River, unlike Hampton Roads, was not a natural deepwater port. Also, the Patapsco received the erosion from a four hundred square mile watershed, most of it hilly and highly erodable, and thus contributed much more silt to the harbor than did the small and slow flowing rivers of the Hampton Roads area.

In the nineteenth century, most of the material dredged from the channels was simply transported to a nearby area and dumped overboard. From the beginning, this practice bothered oystermen, who claimed that it smothered their beds. On the other hand, port authorities complained that oystermen damaged the channels by dredging for oysters along their edges. Such, apparently, was the nineteenth century standoff on the issue.[14]

With the expansion of the port and approach channels in the 1950s, the Corps of Engineers began to dispose of spoil in the Kent Island Deep, an area of deep water that lay to the east of the main channel near the Bay Bridge. This practice was strongly objected to by

Kent Island residents and watermen. Gradually it also became a matter of increasing concern to fisheries biologists and managers.

In 1959, after years of receiving complaints on the issue, Governor Tawes appointed a committee to study the matter. Although the primary concern was the biological effects of spoil disposal, arguments arose from Kent Island property owners that the filling of the Kent Island Deep was changing the currents along the shoreline and greatly increasing the rate of shore erosion. They also claimed that the water along the shoreline was made more turbid by the dumping. Although both of these claims were considered farfetched by scientists, they added to the pressure on politicians to act.[15]

In 1961 the governor's report recommended the formation of a permanent commission to review all disposal questions and other issues involving the use of state submerged lands. It also recommended that the state adopt a policy of land disposal of spoil where possible: "The deep waters of the Bay remain an important ecological environment which should remain free from encroachment whenever possible."[16] The governor promptly formed a Submerged Lands Commission, made up of various state natural resources and economic development interests, and chaired by the state comptroller.

In the mid-1960s, the spoil disposal issues raised by the expansion of the Chesapeake and Delaware Canal took the spotlight away from the Kent Island Deep issue. The canal, which was being widened from 250 to 450 feet, and deepened from 27 to 35 feet, produced an enormous quantity of dredged spoil. Much of that taken from the canal proper was disposed of on land, but the dredging in the approach channels was pumped to nearby shallow areas and to the Pooles Island Deep. The concern of the Maryland fisheries officials was that this spoil would damage the spawning and nursery areas for a number of important finfish, principally the striped bass. It might also put further stress on the salinity-limited oyster beds in the upper Bay, which had been on the decline since 1881.

Significant attention was given to this issue by both the Submerged Lands Commission and the Board of Natural Resources. For a time it was proposed that all dredged spoil be placed on marshlands on Aberdeen Proving Ground. By the mid-1960s, the competing state interests had reached a compromise that would allow for some disposal on land and some in Pooles Island Deep. The Corps of Engineers, however, would not accept the additional costs involved in such a practice. So eventually the state resources agencies had to back down under pressure from port development interests, and the over-board disposal continued for both the Chesapeake and Delaware and the Baltimore Harbor projects.[17]

In the meantime, a number of studies were conducted that tended to diminish concern that the dredging and spoil disposal was having a pronounced adverse effect. The primary result of these studies, aside from advancing knowledge of the biology of the upper Bay, was to provide guidelines for the timing of dredging and spoil disposal, so as not to interfere with spring spawning of striped bass. These guidelines were adopted and the spoil issue appeared for a time to be a problem solved by debate and compromise, and by the application of applied scientific research.

In 1968, however, an additional element entered the spoil disposal picture. Research had been going on for about a decade on the uptake of heavy metals by shellfish. For a shorter time, detailed studies were being made of the water quality and sediments of Baltimore Harbor. In 1968, the Submerged Lands Commission was officially informed of the problems posed by the high concentrations of heavy metals found there.[18] To dispose of these sediments in the open Bay would raise the possibility that shellfish would concentrate these metals to the point that they would pose a threat to human health. At that time, there was substantial uncertainty as to the actual risk involved. Little was known about actual physical, chemical, and biological routes by which heavy metals might find their way into shellfish. Furthermore, there was little agreement as to the levels of metals that should be tolerated by humans.

Despite these uncertainties, the need for action was politically clear. Here was once again a threat to the oyster industry, if not in the direct form of a public relations problem that might affect the marketing of oysters, whether contaminated or not. The state moved quickly to develop a solution. Following a study, the Submerged Lands Commission proposed the construction of a contained spoil disposal area to receive contaminated spoil from Baltimore Harbor. A number of possible sites were considered; that one found most economical and feasible called for the construction of a gigantic man-made island in the vicinity of Hart and Miller islands just to the northeast of the harbor entrance. Back in the 1890s the Baltimore Sewerage Commission proposed to build the Baltimore sewer outfall over these islands. The original plan called for the island to be approximately two miles long, half a mile wide, and twenty-five feet above mean water, with a total capacity of 125,000,000 cubic yards. The plan received the approval of the state resource agencies, the Port Authority, and the Corps of Engineers, all of whom were represented on the Submerged Lands Commission. Here are key words in the proposal, formally adopted by the governor and submitted to the General Assembly in February of 1969:

> Confined disposal of certain types of dredging spoil is essential to the preservation of the environmental integrity of Chesapeake Bay. Provision of the contained disposal area will assure the continued orderly development of water-oriented industry of the State without jeopardizing water quality for at least the next twenty years.[19] The General Assembly quickly approved a bond issue for $13 million to construct the disposal area, and the detailed engineering studies for the project were begun.[20]

With a solution to the contaminated spoil disposal problem in the offing, Maryland resources agencies argued for the suspension of harbor dredging until the Hart-Miller project was completed. Commenting on a Corps of Engineers project in the fall of 1970, the deputy secretary of the Department of Natural Resources recommended disposal, "Since the confined disposal area for the Upper Bay should be in operation by the time this dredging project is to be initiated [in the winter of 1971-72]." In 1971 the director of the Fish and Wildlife Administration likewise recommended disapproval of a Maryland Port Authority project in Dundalk, citing the risk that overboard disposal in the upper Bay would pose to a "developing commercial product, and freshwater clam *Rangia cuneata*." Federal fish and wildlife and water-quality officials also became involved in the process, since all dredging projects needed a certification that they would not contravene existing water quality standards or federal guidelines. What resulted was a series of negotiated compromises that allowed limited dredging while the details of the Hart-Miller proposal were developed, and while the biological questions relating to contaminated spoil were further studied.[21]

The more significant controversy was not between resource and port development officials, or between state and federal levels of authority. The Hart-Miller project, which had been so carefully worked out between competing interests at the state level, and which was seen by the state resource officials as a major environmental victory for protecting the quality of the Bay, was effectively stalled by determined local opposition. Residents of the two peninsulas of land to the west of Hart and Miller islands found the prospect of a large new land mass objectionable. The reasons for the opposition ranged from objections to destruction of the small natural islands to the claims that the man-made island would change the circulation patterns of Back and Middle rivers, and do irreparable damage to the ecology and recreational values of the entire area. Residents objected to the change to the landscape, to the possibility that the disposal area would smell, that it would pollute groundwater, and that birds using it would sicken and die from exposure to the contaminated wastes. They claimed initially that the island would eventually be used as an industrial park, a use which would further degrade the area. The state then gave assurances that the island would eventually become a state park.[22] At the state wetlands hearing in the spring of 1971, several hundred residents turned out to express their opposition to the project, often jeering the numerous state officials that testified on behalf of the project. With the state giving quick approval to the project, the opponents turned their attention to the federal approval process. Joined by boaters and fishermen who use the existing islands and nearby waters for recreation, they launched a campaign of opposition that has effectively stalled the proposal.[23]

The Hart-Miller controversy is of special interest in this narrative because it represents perhaps the most significant water quality issue in the history of the Bay (at least up to 1972), when viewed from the standpoint of its economic impact. Yet, its origins, stemming from concerns for the marketability of Bay oysters, were largely a matter of conjecture. In 1965 or 1970 no evidence existed of an actual health risk. Certainly no modern equivalent of the 1900 Wesleyan incident prompted the blocking of the earlier Hart-Miller project. Nonetheless, the enlargement of the Baltimore Harbor facilities had been effectively stymied. A body of scientific opinion now even suggests that it would be safer to use deep water overboard of contaminated spoil, rather than a contained area, since the material would be returned to the environment in which it had been stable for some time, in which there would be little biological activity, and in which it would not be exposed to oxygen and the leaching effect of rainfall, as it would be in a contained area.

## THERMAL DISCHARGES

If spoil disposal in the upper Bay has been the issue involving the longest conflict between competing Bay uses, the use of the Bay for power plant cooling water was perhaps the issue that developed the most intense controversy. Starting (in 1961) with a coal-fired power plant at Chalk Point on the Patuxent River, then another conventional plant at Morgantown on the Potomac, and a nuclear plant on the James River in Virginia, the issue culminated (in 1971) with the Calvert Cliffs nuclear power plant on the main stem of the Bay in Maryland.

Maryland officials had first become concerned about the question of the biological effects of cooling water discharges while reviewing the proposed power plant on the free-flowing portion of the Potomac at Dickerson. There was an emerging body of literature which recognized that elevated temperatures could alter river biota. When the Chalk Point plant proposal was received in early 1961, a new set of questions were raised because of the estuarine character of the river. The Water Pollution Control Commission therefore established a research program to be conducted by the Natural Resources Institute, a research and teaching arm of the University of Maryland. The major component of NRI became the Chesapeake Biological Laboratory at Solomons.

The concerns of the Maryland officials were threefold. First was the effects of the exposure to high heat on organisms taken into the power plant condensers. Second was the effect of the chlorination that was necessary to keep the condensers free of algae and other fouling organisms. Third, and most important, was the effect of the heated-water discharge on the ecology of the estuary in that area. Since the plant involved a substantial temperature rise, and used a relatively large portion of the available water for cooling, it was at least possible that the plant would have a major, and possibly adverse, effect on aquatic life.

The research progressed over a number of years, first involving collection of background information, then making comparative studies once the plant went into operation. As the results were completed, a generally unfavorable picture began to emerge. The Natural Resources Institute scientists implicated the plant in a number of adverse conditions they observed, most notably a large crab die-off in 1964, a die-off of finfish the same year, and an increased incidence of "green" oysters in the vicinity of the plant which they attributed to copper coming from the plant's condenser tubes.[24]

The conclusions of Natural Resources Institute were generally challenged by officials in the Department of Chesapeake Bay Affairs, who felt the issue of green oysters was exaggerated and threatened to raise a new issue of concern in the minds of the consuming public. Scientists hired by the power company put forward reports that argued that the plant had had little adverse effect. The Department of Water Resources generally accepted the research of the power company, and even later suggested that the objectivity of the Natural Resources Institute group and its work was subject to question.[25] And scientists at the Chesapeake Bay Institute, who were also conducting research on power plant discharges, generally felt that the effects of Chalk Point plant were minor and, for the most part, temporary. Virginia Institute of Marine Science scientists, on the other hand, were generally disposed to take the same view as the Natural Resources Institute scientists, that the power plants posed substantial risks to the Bay environment.

This variance of professional and scientific opinion might have been of little long-term account, had it been restricted to Chalk Point. In fact, however, the results of Natural Resources Institute's work there was being released and discussed in the press during the same period in which Maryland was also evaluating a power plant proposal at Morgantown, when a nuclear power plant was under construction on the James, and when Baltimore Gas and Electric Company announced its intention to build a nuclear power plant on the Bay at Calvert Cliffs. In a short time, thermal pollution became a major new issue, and the basic disagreements over Chalk Point expanded at Morgantown and Calvert Cliffs.

The exact sequence of events relating to the three power plants is now virtually impossible to sort out, particularly since much of the early consideration of the thermal issue went unrecorded. However, the principal initiator of the public controversy was the announcement by Baltimore Gas and Electric in May of 1967 that it intended to build a massive nuclear power plant on Chesapeake Bay. Many different interests opposed this proposal. The first concern was for the cliffs themselves, since the proposed plant was at the site of extraordinary fossil deposits from the Miocene period. Second were the concerns of property owners in the vicinity of the plant, particularly those whose properties would be traversed by the transmission lines carrying the electricity to the Baltimore area. Third were concerns for the risks of radiation, nuclear waste disposal, or accident. And, finally, there were those who concentrated on the possible effects of the waste heat discharged to the Bay.

The Baltimore Gas and Electric Company's Calvert Cliffs nuclear power plant. This aerial photograph shows the entire complex except for the discharge station which is off to the upper right. The circular containment buildings house the reactors. In the foreground is the intake embayment area that draws cooling water from the Bay. Courtesy the Baltimore Gas and Electric Company.

Although all these concerns received considerable attention, and perhaps the land use issue generated the most active opposition, it was the thermal question that received the most attention from officials and the press. By early 1969, enough public concern had been generated to result in six bills being introduced in the General Assembly regarding power plants on Chesapeake Bay, including one that would impose a five year moratorium until a study could be conducted on the effects of waste heat discharges.[26] None of the bills passed, but opposition to Calvert Cliffs continued to grow.

In the meantime, the Department of Water Resources had granted a permit for the operation of Morgantown, an issue that had been under study for over two years. Primarily because of the controversy over Calvert Cliffs, and fed by the reports of NRI from its Chalk Point studies, the Morgantown action also drew heavy fire, some of it from the Virginia side of the river.

Prompted by these objections, the director of the Chesapeake Bay Institute, Dr. Donald Pritchard, issued a statement in which he appraised the likely effects of Morgantown on the Potomac. His reasons for doing so were explicit:

> The considerable newspaper coverage of statements concerning possible excessive damage to the Chesapeake Bay and its tributary estuaries from thermal effects of electric power plants now under construction or planned, and the expression of concern over the members of the State legislature, has led to the general public impression that the Chesapeake Bay and its tributary estuaries are in imminent danger of catastrophic damage to commercial fisheries, sport fisheries, recreation, and other uses. It is my firm opinion that such is not the case. . . . In my opinion, the public as well as their elected representatives have been misled as to the scope of the problem, and the degree to which existing knowledge can be utilized in appraising the problem. . . . A number of scientists and pseudoscientists in this state have been quoted, and misquoted, as predicting severe damage from thermal pollution. Those of us who have contrary views have so far remained publicly silent. I find it necessary to break this silence.[27]

He then went on to appraise the Morgantown project in the light of the Natural Resources Institute findings on Chalk Point, and in the process leveled a direct criticism on the findings of its research and the conclusions drawn from them. In a cover letter to a state senator, he summarized his opinion on the NRI work:

I am of the firm opinion that nowhere in the report published by the Natural Resources Institute . . . is there a demonstration that man's use of and harvest from the waters of the Patuxent estuary adjacent to the Chalk Point Power Plant has been adversely effected [sic] by excess temperatures resulting from operations of that plant.

He ended his report with the unequivocal statement:

I must conclude that the operation of the Morgantown Power plant, under conditions prescribed by the Department of Water Resources, will have no measurably effected [sic].[28]

Dr. Pritchard's statement was quickly followed by one from the Director of the Natural Resources Institute, Dr. L. Eugene Cronin. In it, Dr. Cronin made a point-by-point response to the issues raised by the Chalk Point study. Then he summarized his major concerns to which "the biologists of the Virginia Institute of Marine Science and members of the Potomac River Fisheries Commission concur[red]":

a. Allow no elevation above 90 degrees F.; b. Prohibit the use of "tempering water" [a practice of drawing additional river water into the discharge canal so that the final temperature would not exceed the prescribed maximum]; and, c. *The biological studies by company-hired consultants are poorly designed, contain serious errors and are completely inadequate as a baseline of present conditions or for predictions of future effects.*[29] (emphasis supplied in original)

That Dr. Cronin saw fit to underline the last point makes it clear that this was a major source of concern and disagreement.

He then went on to characterize his difference with Dr. Pritchard:

You and I differ principally in philosophy, and this should not be misconstrued. You are inclined to encourage full use of the Bay unless available knowledge proves that human uses will be impaired. I am more conservative and prefer not to risk damage to the Bay until reasonably good estimates can be made of all effects.[30]

He concluded by quoting Dr. Pritchard's final statement to the effect that there would be no adverse effect from Morgantown (quoted above), and commenting on it thus:

With the best will in the world, I must say that the fragmentary information available is totally insufficient as a basis for such an

absolute statement, which does not sound like your usual pre-
cise and well-qualified thinking. . . . Perhaps it is easier for
physical oceanographers to make such flat predictions than for
biologists. I would never make them, and feel that it is of the
utmost importance that we all avoid sweeping conclusions.[31]

These two statements, issued by the two leading Maryland scientific policy advisors, provide a striking outline of the type of problem faced by Bay officials since about 1960. Dr. Pritchard laid particular stress on the public relations aspects of the controversy. Dr.
Cronin stated a basic difference in point of view ("philosophy" in his
words) that would lead to very different government policies and
actions, depending on which was adopted. And behind the debate was
the question of the reliability and objectivity of the scientific findings
developed by various laboratories and individual researchers.

Faced with these unresolved questions, Governor Mandel, at the
request of the General Assembly, appointed a commission to study
the matter. Under the chairmanship of Dr. William W. Eaton, the
committee examined the record both as to the thermal and nuclear
issues. In late December, 1969, it issued its report, saying in effect
that the Calvert Cliffs plant, if built under the standards imposed by
the Department of Water Resources and the Department of Health
and Mental Hygiene, would not pose a threat to the Bay.[32]

The commission was clearly impressed by the economic argument that the Bay represented a valuable resource as a cooling medi-
um. It pointed out that the entire Maryland portion of the Bay would
only be elevated in temperature by 1.5 degrees F. by thirty power
plants the size of Calvert Cliffs (assuming even distribution of the
waste heat), and it concluded therefrom: "Research to find what
fraction of the potential several hundreds of millions of dollars cost of
artificial cooling systems can be avoided through Bay use hardly
needs economic justification."[33] In March of 1970, Dr. Cronin and the
principal investigator at Chalk Point, Dr. Joseph A. Mihursky, issued
a press release in which they itemized a number of unanswered
questions regarding the effects of the plant, and suggested that a
strong research program be adopted to address those questions. If the
plant is approved, they said, and is proven to have adverse effects,
penalties should be imposed and immediate corrective action
taken.[34]

The document also included additional statements by Drs. Cronin and Pritchard. The former stated his basic precept:

It is of the utmost importance that the effects of large environmental changes be estimated with reasonable accuracy prior to decision and commitment on those changes.[35]

The latter commented:

It is an unfortunate fact that much of the research on the effects of thermal discharges into natural bodies have [sic] been biased, even though possibly unconsciously, towards either showing that no damage occurs, or is intended to prove the conservationist's opinion that catastrophic damage will result from such discharges. While it will be difficult to assure that all research personnel are free from some bias in this matter, particularly in view of recent events here in Maryland, steps could be taken to assure an unbiased review and research guidance procedure.[36]

In May 1970, the Department of Water Resources issued its final permit to the Baltimore Gas and Electric Company, which allowed the appropriation of roughly 3.5 billion gallons of water per day for passage through the plant as cooling water. This permit disposed of the principal water-quality hurdle at the state level, and was followed in 1971 by a Certificate of Public Convenience issued by the Maryland Public Service Commission, a write-off by the United States Environmental Protection Agency to the Atomic Energy Commission, a Maryland Wetlands Permit, and eventual approval by the Atomic Energy Commission itself. (It should not be thought that approvals were straightforward and without difficulty. The numerous legal and administrative issues involved in this project would take a book length manuscript to detail, including the landmark decision by the United States District Court of Appeals in July 1971, that found the Atomic Energy Commission inadequate in its fulfillment of the requirements of the National Environmental Policy Act. The details surrounding the various permits issued by Maryland alone occupy several file drawers.) The Calvert Cliffs plant began operation in 1974, and in 1976 produced more electricity than any other nuclear power plant in the free world.[37] In 1980, the company claimed that $400 million had been saved as compared to generating the same amount of electricity using oil as fuel.[38]

While this major conflict was erupting in Maryland, Virginia went about the business of approving a nuclear power plant on the James River at Hog Island with little fanfare and no visible controversy. In the words of the director of the Water Pollution Control

Board, it "went through as slick as a whistle."[39] The absence of controversy itself presents a problem in analysis, because there is much less in the written record from which to judge why things were so different south of the Potomac. The Virginia Fisheries Commission did comment in a general way that thermal pollution was an additional threat to the Bay, but there was no overt attention to the Surry Plant. One fact that no doubt contributed to the smoother approval process was that the hydraulic model for the James was available to test the physical effects of waste heat discharge. Working under contract with the Virginia Electric Power Company, scientists were apparently able to satisfy Virginia government officials that these thermal effects would not have adverse biological consequences. That the whole process in Virginia occurred several years before the Morgantown and Calvert Cliffs issues in Maryland probably made for a smoother approval. No in-state research had alleged adverse effects of other power plants: the Chalk Point work in Maryland was just emerging as the Surry Plant was going through its final approvals in 1967. Also, the state permit process was significantly less cumbersome in Virginia, so that decisions were more focused both in place and time. Perhaps most significantly, there was no concentrated neighborhood or conservation group opposition, as in Maryland, and little comment on the project by the press. By the time the Bay-wide press began to treat wastes as a major issue, the Surry Plant had already been approved and was well along to completion. Thus, the issue that perhaps steamed the tempers of more persons in Maryland, if not the waters of the Bay, caused nary a mild fever in Virginia.

## INTERSTATE ISSUES

During the 1960s, the interstate management of Potomac and Susquehanna rivers came under consideration. The states of New York, Pennsylvania, and Maryland joined together to form a Susquehanna River Basin Advisory Commission. Their task was to develop a compact mechanism for addressing the management of interstate issues.

Because very little of the Susquehanna River is in Maryland, officials there were chiefly interested in addressing the relation of the river to the Chesapeake Bay. The tremendous freshwater inflow of the river was recognized as the controlling influence on the salinity and long-term circulation of the upper Bay. Moreover, the Bay receives the wastes carried by the Susquehanna to its mouth. Maryland sought to use the proposed compact to address its interests in both the quantity and quality of Susquehanna River flows.

In the mid-1960s, various officials of Maryland resources agencies participated in discussions considering regulation of flows in the Susquehanna. They took the same stand that Virginians had taken about the James. They gave general support for a regimen that would reduce flooding and increase minimum flows. Such an approach would tend to increase the range of the oyster while reducing the range of its principal enemies. In addition, reduction in spring floods would reduce the bacterial contamination that every year required the closing of the upper Bay to shellfish harvesting in the spring months.

Much of the interest in the Susquehanna in the early 1970s arose from the severe drought experienced by the eastern United States from about 1961 to 1966. The Susquehanna, as the largest eastern river discharging to the Atlantic, was looked upon as a potential water supply not only within the basin, but also in the adjacent Delaware and Potomac basins. During the same period, greatly increased use of river water for irrigation was being developed, and power plants were being planned that would require substantial amounts of water for cooling.

By the end of the decade, these issues were being eclipsed by concern over the diversion of fresh water from the upper Chesapeake to the Delaware Bay by use of the Chesapeake and Delaware Canal. The issue began to take form in January, 1968, when Dr. Pritchard, while working on a Delaware estuary problem, discovered a Corps of Engineers study done in 1934 that estimated that there was a net outflow from the Chesapeake to the Delaware Bay of slightly less than one thousand cubic feet per second. Since the canal was being enlarged from 250 by 27 feet to 450 by 35 feet, Dr. Pritchard reasoned that the net outflow would substantially increase. He saw this as a factor perhaps more significant to the physical and biological properties of the Bay than the other diversions of Susquehanna flow then under consideration. He made this possibility known to Maryland officials and other scientists, and began his own calculations on the probable effect of the canal enlargement.[40]

The issue was quickly picked up by government staff workers, citizens conservation groups, watermen, and the press. By early 1969, the Chesapeake and Delaware Canal had joined the list of major Bay problems, along with Calvert Cliffs, wetlands destruction, and spoil

disposal, in addition to the usual concern for municipal and industrial discharges. Expressions of concern ranged from the carefully
couched statements of Dr. Cronin, who said that the effects could not
be predicted, but might well be significant, to the plaint of a waterman who suggestd that the canal be made big enough so that instead
of having a dead bay (Delaware) connected to a half-dead one (Chesapeake), we could have just one big dead bay.[41]

In 1970, at the urging of members of the Maryland Congressional
delegation, the Committee on Public Works of the House of Representatives held hearings on the matter. Ostensibly the committee
was to consider whether the project, which had been authorized in
1954 and which had already been largely completed at a cost of $65
million, should be continued.[42]

Testimony was varied. The committee was considerably disturbed by the recommendation of a senior Department of Interior
official that the project be stopped until exhaustive oceanographic
and biological studies could be performed. Members expressed their
dismay that a federal agency would come forward at such a late date
and make a case against a project that had been available to public and
official scrutiny for so long. The Interior Department official responded that new information justified taking a fresh look at a large-
scale project of this kind, and that the value of the Chesapeake Bay
was so great that a cautious approach was called for.[43]

The key testimony came from Drs. Pritchard and Cronin, who
issued statements arguing that the probable effects of completion of
the project would be small, but that studies should be conducted that
would provide adequate physical and biological understanding of the
Bay so that the long-term effects of the enlargement could be as-
sessed. They suggested that such studies would indicate if there was a
need for corrective measures (such as locks), if adverse biological
effects could be attributed to the canal.[44]

Following a recess of several weeks, the hearing reconvened with
testimony from the Corps of Engineers and Drs. Pritchard and Cronin, outlining a research program to address the issues raised previously. The committee responded favorably to the testimony, with
the general sense that the project could proceed with concurrent
research. The possibility that research results might indicate the
need for locks was again raised, but not addressed in any detail.[45]

Three months later, a research contract was awarded to the University of Maryland of $798,600 to conduct studies of the upper Bay.[46]

That work was conducted over the next several years, as well as mathematical modeling work by the Corps of Engineers to refine the estimates of the physical response of the system to the enlargement of the canal. By the end of this study period (1972), the research was well along, the plug in the western end of the canal still in place (due to a lack of funds and delays related to the spoil disposal issue), and the Chesapeake and Delaware Canal was essentially a dead issue as far as Bay water-quality discussions were concerned. Thus, in the space of just a few years, an issue was defined on the basis of initially obscure and largely technical data, was elevated through official and journalistic attention to the status of a crisis, and passed quickly and quietly to oblivion, to join such former issues as Eurasian milfoil and explosives testing as items removed from the Bay political agenda.

## COMPREHENSIVE POINT OF VIEW

The 1960s marked the beginning of systematic attempts to address Bay governance from a comprehensive point of view. The 1933 Chesapeake Bay Authority had been the first such attempt, but it left little evidence except the document that records the conference. Beginning in 1960, however, both states and several federal agencies conducted studies or planning efforts that purported to be comprehensive, and in 1971 there was a brief attempt to address the interstate aspects of Bay governance. Bay water quality matters, although not the only element of these efforts, was certainly the dominant one, and was linked to most of the others.

The first activity was a study of water resources management in Maryland by the Board of Natural Resources. Although not specific to the Bay, it addressed a number of Bay issues, and made as its summary finding a recommendation that water management be structured at the state level so that the relations between various functional areas could be addressed.[47] This study was followed by a major study of water management in Maryland that led in 1964 to a wholesale reorganization of the major natural resources agencies in the state. In particular with regard to the Bay, the Department of Tidewater Fisheries was reconstituted as the Department of Chesapeake Bay Affairs, with the responsibility for developing a comprehensive plan for the management of the Bay, together with its more traditional duties of managing the sport and commercial fisheries of tidewater Maryland.[48]

Also during the early part of the decade, the Public Health Service was developing a basinwide perspective on the Susquehanna-Chesapeake system. Starting in 1963, it established a formal Chesapeake Bay-Susquehanna River Basin Project, which was scheduled to run for six years and involve a staff of up to seventy-five persons.[49] The stated purpose of the project was to perform basic water-quality studies and develop a basinwide framework within which water quality and other management decisions could be made. The study staff saw its mission to bring into sharper focus the relation of the free-flowing river to the Bay.[50] Of particular interest was the question of enrichment of the Bay, which some of the researchers considered to be the most important water-quality problem for the Chesapeake.

After an ambitious start, the project became the victim of changing priorities within the Federal Water Pollution Control Administration, which came into being in 1965 and took over the water quality functions of the Public Health Service. The project completed a number of specific studies, but never published anything approaching a comprehensive discussion of the basin.

Meanwhile, the numerous political issues on the Bay that involved questions of the effects of engineering changes raised the idea that a physical model for the entire Bay would be a useful research tool. Various members of the Maryland Congressional delegation particularly were impressed by the range of Bay problems and the need to address them in an integrated way. In 1965, the Congress authorized the Corps of Engineers

> to make a complete investigation and study of water utilization and control of Chesapeake Bay Basin . . . including . . . navigation, fisheries, flood control, control of noxious waste, water pollution, water-quality control, beach erosion, and recreation.

To aid in this study, the Corps was authorized to build a hydraulic model of the Bay.[51] The resulting study covered a span of approximately ten years, and led to the publications of the multivolume *Existing Conditions Report* and *Future Conditions Report*,[52] representing by far the heftiest, if not the most illuminating, discussion of the complex relations between various Bay functional areas. The model was constructed in Matapeake, Maryland, and has seen limited use as a research tool.

Several other federal studies focused on the Bay in the latter part of the 1960s. Both the National Estuarine Pollution Study of the

Federal Water Pollution Control Administration[53] and the National Estuarine Study of the Fish and Wildlife Service[54] had special sections on Chesapeake Bay where they discussed the complex interfunctional and interstate aspect of the various water quality problems then identified. The National Council on Marine Resources and Engineering Development had a task group that made a case study of the Bay, in order to formulate policy for the coastal zone.[55] And NASA, eager to demonstrate its earth-resources capabilities, established several projects and held a symposium on the applications of space technology to the management of the Bay.[56]

The states were also active during the latter part of the decade. Maryland held a Governor's Conference on Chesapeake Bay in September of 1968 for the purpose of laying "a foundation for the orderly development of Chesapeake Bay."[57] That conference was followed in 1969 by the formation of an intrastate planning body made up of the heads of most of the state agencies having responsibilities affecting the Bay. This body, whose function was made largely obsolete by the major restructuring of state government in 1969 and 1970, was known as the Chesapeake Bay Interagency Planning Committee, and its activities resulted in the publication of a planning document.[58] In Virginia, similar focus was being placed on the James River, where at least two studies and numerous shorter publications of the Virginia Institute of Marine Science attempted to state the resource problems of the river, the relations between various uses, and the research and management needs posed by the increasing complexity and volume of use.[59]

Seen from the perspective of over a decade, these planning and study activities seem complex and intertwined. When considered in relation to the large number of important policy issues and controversies that were being addressed during that same period, they are nothing short of mind boggling. Their principal contributions are doubtful in the ways in which they influenced legislation, public attitudes, and management practices that have since been adopted.

Not only did the decade of the 1960s see a vast increase in the size and complexity of water quality issues; it also saw the dividing line between water and land use issues blur and occasionally disappear. The relation between land activities and water quality took several forms. Most directly, this period saw the general acceptance of, and first action on, the role of non-point source pollution on the quality of the Bay. Second, it saw some recognition that general regional development had direct impacts on water quality. Third, this era produced a number of controversial projects that were opposed on water quality grounds, but at root were conflicts over land use. And, finally, there developed a special concern for the role of wetlands in the physical and biological quality of the Bay.

That land runoff was a potential pollutant had long been recognized in the Bay region. Sediment was acknowledged as a cause in the reduced run of anadromous fish well before the twentieth century. Runoff from farmland was identified as a potential source of bacterial pollution by the Baltimore Sewerage Commission in 1897. The silting over of oyster bars as a result of heavy soil erosion was at least mentioned in the nineteenth century, if not greatly worried about. But it was after 1960 that government officials and institutions began to address the issue directly, beginning with soil erosion in Maryland.

Maryland state officials had discussed the impact of soil erosion and sedimentation prior to 1960,[60] and the Water Pollution Control Commission had concentrated on a specific aspect of soil erosion, that arising from sand and gravel operations, since the late 1940s. Special attention to both these problems was given in the Patuxent watershed. The board began to worry about the heavy development of housing in the Bowie area, where it was announced that a large new community was to be developed in a previously agricultural area. This site was also on the same stretch of river that had a number of sand and gravel operations, and was not very far from tidewater, where there had long been concern for the gradually dminishing depth of water in the upper estuary. By the mid-1960s, with substantial interest from a number of legislators and local officials, the state developed a sediment control act for the Patuxent River watershed.[61] This act required local adoption of soil containment practices for large construction projects, backed by state enforcement powers and technical assistance. This program was adopted and successfully implemented on the Patuxent, and was extended to the entire state in 1969.[62]

Both Maryland and Virginia also faced a relatively new problem during this period with the advent on a large scale of livestock herds and feedlots near tidewater. This enterprise was considerably more of a problem in Virginia, where the concentration of hogs and cattle was higher. Large-scale feedlot operations were, in general, required to contain, if not treat, the waste from their operations, but extensive areas in Virginia were closed to shellfish harvesting at least partly as the result of agricultural land runoff. One of the difficulties of dealing with this problem was the uncertainty of causes of bacterial contamination. In a rural area there might be several sources contributing to excessive fecal coliforms: the discharge from the nearest sewage treatment plant; discharge from septic tanks; bacteria from wild animals, such as muskrats and geese; farm animals; and boats, particularly when concentrated in marinas. The difficulty of pinpointing sources, particularly over time, makes it difficult, both politically and technically, to require remedial action by one sector, even though it might be the most suspect. Runoff in the rural areas of Virginia has thus required the closure of considerable areas to shellfish harvest, without any clear prospect for corrective action.

Few areas in the country experienced more rapid growth than the Washington metropolitan in the 1960s. One consequence was an attempt by local governments to control both the rate, character, and environmental impact of that growth. A variety of means were available: conventional zoning, performance standards on new construction, and extension of public services, particularly roads, water, and sewers. In addition, the unintentional but pervasive sewer moratoria that prevailed had the effect of curtailing or redirecting growth. Many of the reasons for growth management extended well beyond environmental concerns, but protection of streams and enhancement of the Potomac estuary were also recognized as purposes for which storm water management, sediment control, and open space protection were justified. Although the specific programs adopted by the several major jurisdictions are too numerous and complex to itemize, it is especially pertinent to note that local jurisdictions were recognizing that the heavy costs that were going to be required to bear in achieving advanced waste treatment of sewered wastes would perhaps be in vain if corresponding reductions were not also made in the waste loadings from non-point sources. This issue came under formal consideration on a regional basis under the aegis of the watershed planning provisions of the Federal Water Pollution Control Act of 1972.[63]

It is a truism of environmental regulation that development projects are often opposed on environmental grounds when the actual reasons for the objection lie elsewhere. Proof of this in a particular instance is difficult, because it requires an inquiry into motivations of those who oppose a particular project. This difficulty is well illustrated by the early experiences of the Maryland Wetlands Act: in case after case a tidewater construction project would be opposed for reasons having nothing to do with the intent of the act. In general, opposition of small projects, such as bulkheads and piers, was based on disputes over riparian rights or questions of property boundaries. In many instances, opponents of a particular project did not even bother to address the wetlands protection issues involved.[64]

In the two cases that follow, the judgment that the opposition was arguing primarily over land use rather than water quality is that of the authors. Although the record is clear, at least, that the issues were mixed, these cases illustrate notwithstanding that non-water quality issues became a factor in decisions that were reviewed or decided primarily on the basis of water quality and related environmental considerations. Steuart Petroleum and Harry Lundeberg School of Seamanship, the two cases, separate physically and legally, took place in the same neighborhood at about the same time, and were opposed by essentially the same local interests and by the Chesapeake Bay Foundation. They illustrate a common principle—that local opposition based primarily on land use issues can have a definite and perhaps decisive effect when expressed through government involvement based on water quality.

Steuart Petroleum operated an oil handling and shipping facility on the Potomac River in lower St. Mary's County, the primary function of which was to supply the Washington metropolitan area with petroleum products. Ocean draft ships would unload at the facility, and the oil or other petroleum products would be transferred to barges or other shallow draft craft for transport up the Potomac. In the late 1960s, Steuart announced its intention to apply for status as a free port, which would allow it to import foreign crude oil duty free. What was an oil handling facility would become a refinery.[65]

Local opposition was swift and intense. A Potomac River protective association was formed, and opposition was focused on the requirement of a building permit from the county commissioners before work could proceed. Opponents concentrated on the commissioners and alleged that the tax advantages of a major new facility, which was a prime attraction to a relatively poor rural government, would be more than offset by losses suffered by area watermen, recreation-oriented businesses, and by property owners due to the inevitable water quality degradation that the oil facility would cause. Much was made of numerous oil spills that had occurred at the facility. Testimony from watermen argued that catch had declined in the area due to oil pollution. Despite the company's claims that it would operate an environmentally safe facility (and despite the presence of a Coast Guard station in the shadow of the facility, thereby making it virtually inevitable that any violations of pollution statutes would be detected and prosecuted), the commission denied the necessary county permits. The company thereafter withdrew its application for free trade port status, and abandoned its plans. (Another refinery proposal was advanced by Steuart in 1973, which was also successfully opposed by local interests.)

The press labeled it as an example of citizen concern for clean water.[66] It is arguable, however, that water was not the prime issue. The threat to local water quality was nebulous and unsupportable in advance of the construction. What was clear was that the facility would represent a major industrial intrusion, in this case an expansion, into an area notable for its isolation and natural charm.

On the other side of the narrow peninsula from the Steuart facility, St. George's Creek was being attacked by another intruder in the form of the Harry Lundeberg School of Seamanship. The school was operated by a small but politically influential union to train young people for maritime trades. In order to provide adequate draft and berthing space for the many vessels used by the school, the union embarked on a massive dredging campaign. Under permits from the Department of Water Resources, it operated its own hydraulic dredge, day and night, to dredge several hundred thousand yards of material, which it deposited within a large upland diked area across the creek from the school.[67]

Intense local opposition to the project alleged that it produced heavy sedimentation of oyster beds, that there had been oil spills from the dredge, and that the school was exceeding and breaking its permit restrictions in various ways. Neighbors also objected, understandably, to the night operations of the dredge, which made considerable noise. As a result of these complaints, various additional conditions were placed on the operations, and it was subject to several temporary shutdowns for violations.

In addition to the main dredging near the school, a permit had also been applied for to dredge a small cove on which the school owned other property, across and up the creek. With the passage of the Maryland Wetlands Law in 1970, a hearing on this project was held in May 1971. Local objectors to the project were numerous and passionate. They presented a litany of offenses and adverse consequences of the school's other dredging activities, claiming above all that the school was damaging the fisheries resources of the creek through siltation, destruction of the natural bottom, and oil and other wastes from the dredge and other vessels. Investigation by various technical specialists from the Department of Natural Resources suggested that these claims were probably not well founded, and that if the creek was undergoing environmental change it was due to a large number of factors, such as development in the watershed and along the shore-line, which was having a greater cumulative effect on the creek than would the relatively localized dredging by the school. Nonetheless, the license was denied, on the basis that the wetlands act established a presumption that undisturbed wetlands (in this case submerged lands) were in the public interest, and anyone wishing to alter or destroy wetlands had to make a case that there were offsetting bene-fits to be derived from the project. In this instance, the school was found not to have made a sufficiently strong case that the project was beneficial. Indeed, it appeared to the hearing officer that the school's primary objective was to convert a shallow cove into a bulkheaded harbor simply out of a seaman's preference for deep water and a firm shoreline.

What emerged from the testimony of many opponents of the project was that a primary concern was the very presence of the school itself. The school, like Steuart, was an intruder that threat-ened to alter the character of the area for the longtime residents and the retirees that enjoyed its isolation and natural beauty. The school portended continuing physical changes to the area, and brought large numbers of strangers into an area that had been socially insular and stable. Certainly there were water-quality issues involved with both projects, but the overriding concern was over more fundamental social, economic, and aesthetic changes that these two neighboring facilities represented. The water quality and wetlands statutes ad-ministered by the state simply provided the most visible means for opposing them.

Concern for wetlands protection became a national environmen-tal issue in the 1960s. Numerous eastern states were enacting wet-lands protection laws in the late 1960s and early 1970s. On the Bay, with its extensive tidal marshlands and shallow submerged lands, Maryland enacted such a law in 1970[68] and Virginia in 1972.[69] Wet-lands were recognized as important in the life cycles of waterfowl, wildlife, fish, and shellfish, as mechanical buffers against erosion and

Marshland near Cove Point, Maryland, 1972. Preservation of sites similar to this has become a legislative issue. Since the photograph was taken, the Columbia Gas Company has occupied this area for its liquefied natural gas facility. Photograph by Richard McLean.

tidal flooding, as sediment traps, and as sources of beneficial organic matter and nutrients for aquatic life. In the last two categories, particularly, wetlands were seen as having a beneficial water quality role. (In Maryland there was also a major impetus to prevent land developers from enriching themselves by creating new land out of the public domain by dredging and filling. Here also, wetlands projects had land use implications that were often of more public concern than the impact on the aquatic environment.)

Although the Virginia statute was not implemented until after the closing date of this study, the Maryland statute operated for approximately two years. Perhaps its principal interest lies in its providing for the first time a state law that required a broad examination of environmental issues relating to tidewater development. A project was to be examined based on the "ecological, economic, recreational, developmental, aesthetic, and agricultural values" it affected. This list is admittedly broad and general: the act gave no guidance as to what these terms meant or how they were to be weighed and balanced. Nonetheless, most tidewater development projects were brought under a broader review than they had previously been. The wetlands process was also tied indirectly to a more specific water-quality evaluation, based on a complex linkage with federal statutes. All activities requiring a state wetlands license also required a permit from the Corps of Engineers under the provisions of the Rivers and Harbors Act of 1899. This act, in turn, was linked to the Federal Water Pollution Control Act by administrative agreement between the Corps and the Federal Water Pollution Control Administration (later EPA). The Corps would not issue its permit without state approval, including a certification from the state water control agency that water quality standards would not be contravened by the project.

Thus, in a crude and largely unintentional way, tidewater projects came under a form of state-federal review that included environmentally based land use analysis (wetlands considerations), protection of navigation (the historical purpose of the Rivers and Harbors Act), and water-quality concerns (through the linkage to the Federal Water Pollution Control Act). Although as a statutory and administrative framework, it proved inadequate to deal with the complexities of issues like Calvert Cliffs and Hart-Miller islands, it at least represented a start in bringing together public consideration of the wide range of water quality and related environmental issues that the controversies of this era demanded. Clearly what was needed was not a static comprehensive plan for the Bay, which had been a popular cry of the late 1960s, but rather a process whereby the various and often competing interests of the Bay could be evaluated and balanced, based on some broad but reasonably explicit public policies. In virtually all cases, at least the starting question has been, and will continue to be, "What effect will this project have on the water quality of Chesapeake Bay?"

# TWELVE

# Perceptions of Chesapeake Bay

[It is] a fair Bay compassed but for the mouth with fruitful
and delightsome land. Heaven and earth never agreed better to frame a place for man's habitation.
                                    —Captain John Smith (1608)

The Bay is dead.
                                    —Anonymous (1970)

The popular perception of Chesapeake Bay has changed. This change
is manifested in the journalistic accounts of the Bay. Originally the
Bay was viewed as a gift available to serve man's pleasure. Of late it
has become an object of animistic concern possessed of a natural life
and an indwelling soul. Newspaper headlines ask, "Is the Bay Dead?"
while the Chesapeake Bay Foundation exclaims, "Save the Bay!" The
story of how the press, public opinion, and politics have interacted to
change the Bay from an inexhaustible resource to a finite and endangered one affords a good summary of our Bay history.

Early publicity treated the Bay as a bounty. On into the nineteenth century articles abounded that described the delights of the
Chesapeake, the abundance of its resources, and the ruggedness of its
watermen. However, in the course of this history there was the bad
news which began to change the Bay in the public's perception. By the
end of the nineteenth century the germ theory had been accepted and
with it the recognition of the Bay's "disease-engendering and pestilential conditions."[1] At the same time Bay cities were developing
municipal water supplies and sewer systems to carry off wastes from
growing residential and business districts. Bay tributaries adjacent to
the tidewater towns became the sink into which these vast quantities
of liquid were emptied.

Except for its effects on oyster culture, the use of the Bay as a
disposal area seemed a happy and inexpensive solution to urban

waste problems. At the turn of the twentieth century, when scientists indicted the oysters as a carrier of typhoid fever, the Bay oyster industry faced the marketing problem of how to quiet the concerns of the "pure food faddists." One response was the construction in Baltimore of a sewerage purification plant; when placed in service on Back River in 1914 it became this country's first major treatment facility. The lobbyists who supported sewage treatment were more intent on cleansing the reputation of Maryland oysters than on cleansing the waters of Chesapeake Bay.

The "oyster scare" publicity was not so easily dealt with. Epidemics in the mid-1920s, which were linked to raw oysters, led to water-quality and processing standards for oysters overseen by the United States Health Service. Maryland watermen continued to act as a clean water lobby. Maryland reduced oyster bed closures with a program of sewage treatment plant construction that led the nation. Virginia watermen were less powerful. Municipal wastes from tidewater towns poured untreated into receiving water. Virginia's legislature and courts refused to respond to the problem, and widespread oyster bed closures resulted.

One reason why Maryland's oystermen proved more politically powerful than those of Virginia was that there were more of them. By 1900 Virginia had embraced a private oyster culture. As a result, Virginia's oyster fishery was dominated by a relatively few leaseholders with larger, more efficient operations than in Maryland, where oyster fishery remained open to all. The large number of Maryland oystermen comprised a powerful lobby. Unfortunately, their large number was also destroying the natural oyster bars through overfishing. Despite editorials and feature articles lamenting the decline of a once-great industry with titles such as "The Vanishing Oyster"[2] and "Oyster Stew,"[3] the legislature refused to limit access. In this context, at least, the political clout of Maryland oystermen, intent on preserving the status quo, proved greater than the persuasive power of the press.

While the Maryland oyster interests created an effective lobby for clean water, their interests—bacterial discharges which polluted oyster beds—were specialized and localized. Sport fishermen joined them in support of clean water, but their interests related primarily to pollution of small free-flowing streams rather than pollution in the Bay. Meanwhile in port areas, oil was dumped from the bilges of vessels plying the Bay and mixed with the ubiquitous discharges from

Modern Baltimore, taken from the top of the State Office Building. On the horizon is the tidal Patapsco River and, to the left, the Francis Scott Key Bridge. By 1980 when this photograph was taken, Baltimore had reigned as metropolis of the Chesapeake for one hundred and fifty years. Its dominance is suggested by the mountain range of buildings built on the original sixty acres of Baltimore Town, laid out in 1739. Photograph by Richard McLean.

food-processing plants. In addition, industrial discharges in the Curtis Bay area of Baltimore and portions of the York River attracted some attention, but in the absence of a fish kill, little effective action resulted. During the first half of the twentieth century no group argued rigorously the case for good water quality throughout the Bay.

The post-World War II years in the scientific and technical work advanced understanding of water quality issues. By the early 1960s new pollution concerns joined the long-standing attention given to municipal and industrial wastes. Concern for the effects of sediments and nutrients directed attention to activities throughout the Bay watershed. Problems associated with the disposal of dredged material became a matter of intense interest, particularly in Maryland. By the mid-1960s, waste heat from electric generating plants became a prime concern in Maryland. Also, at about the same time, controversies arose in both states over a number of engineering projects that had the potential to alter the circulation and salinity patterns of biologically significant areas in the Bay. Throughout the 1960s, the Washington metropolitan area wrestled with the problem of how to dispose of the wastes of a rapidly growing metropolitan area, while at the same time restoring the quality of the badly polluted upper Potomac estuary so that it could be used for recreation.

Biologists and physical scientists then became active in Bay affairs. They were called upon to study fish kills and declines in species and to project the likely effects brought on by public works projects, new industry, or waste disposal plants. Scientists also played key roles on boards and commissions that make policy recommendations to politicians and the public. They sought after, and received, public funds for research and for monitoring public programs. Finally, they made economic, political, administrative, and legal decisions.

Scientists became advocates of clean water. In particular cases they were joined by special stakeholders. Since these intervenors— opposed to a transmission right-of-way or diked spoil disposal site— had the good public relations sense to generalize their concerns, to argue that a new power plant would result in thermal pollution or that a dike failure could result in widespread dispersal of heavy metals. It was the scientists, however, who constituted the steady force in the clean water lobby.

The results of their efforts were soon apparent. In the early 1960s, a new kind of article came to be written which raised questions about the health of the Bay and discussed in varying degrees of detail the problems facing Bay management. A sample of titles reflects the tone and content of these articles: "Chesapeake at Bay," "Does Pollution Threaten the Bay?," "The Bay on Borrowed Time," "Our Beleagured Bay," "Are They Killing Our Bay?," "The Bay: An Abused Treasure," "Great Ugly Changes Mar Chesapeake," and "Changes Imperil Bay Resources."[4]

These articles were remarkably similar. Almost without exception, they spoke of the once boundless treasures of the Bay as reported by early settlers. They detailed the abundance of the Bay's resources, particularly the great oyster harvests of the late nineteenth century. They recounted the declines of commercial and sport species of fish and shellfish; they itemized specific threats to the Bay in the form of projects such as Calvert Cliffs or the Chesapeake and Delaware Canal, as well as more generalized threats such as increasing urbanization and industrialization. They quoted Bay scientists and administrators, and they invariably made reference to the "health" of the Bay, often comparing it in some way with reputedly "dead" bodies of water such as Lake Erie, Lake Michigan, or Delaware Bay. They ended with a pessimistic prognosis for the Bay.

As indicated, these articles had shifted attention to a concern for the overall health of the Bay. They reflected the fact that the scientific community had become enthralled with ecology and with its emphasis on the relationship between organisms and their environment. Scientists raised the public's consciousness that the Bay was a single system. They were joined by new allies—citizen environmentalists concerned with the overall productivity of the Chesapeake Bay ecosystem. In 1970 this perception found expression in the political arena when Maryland's Governor Marvin Mandel said:

> We feel an almost sacred obligation to the Chesapeake Bay and its tributaries . . . There is concern for the health of man and some parochial interests of those whose lives will be distributed, but the predominant concern is for the Bay itself, the Bay as a living entity.[5]

The changing perception of the Bay was also reflected in the agencies called upon to govern it. Sheriffs were the first governmental agents involved with the Bay. Legislation in both states empowered these local officials to regulate the harvesting of fish and to enjoin the fishing of streams. With the "oyster wars" of the 1860s, Maryland and Virginia attempted through force of arms, on occasion, to keep the

Bay's bounty for themselves. As fisheries took on greater economic significance in the late nineteenth century, both states formed boards to study, advise on, and eventually manage marine fisheries.

These boards evolved into the Virginia Marine Resources Commission and the Maryland Tidewater Administration. This evolution has been relatively simple in Virginia, with only one simple name change in the last eighty years. In Maryland, constant legislative experimentation and organizational change produced at least ten different entities responsible for the management of tidewater fisheries resources. Despite these changes, attention to the needs of the sport and commercial fisheries appears to have remained.

In the latter part of the nineteenth century, both states formed boards of health which evolved into major state departments. These have retained responsibility for municipal sewage disposal and oyster sanitation. In Maryland, the state health agency has been a major factor in Bay management since the early 1900s for forty years, it regulated both industrial and municipal waste treatment, with support from the various conservation agencies.

In 1939 Virginia created a regional sanitary authority for Hampton Roads, and by 1948 both states had statewide pollution control agencies. Virginia's Water Control Board had primary responsibility for both municipal and industrial waste control; in Maryland that responsibility was split between the Health Department and the Water Pollution Control Commission.

The addition of Federal agencies began with the Corps of Engineers which has had a continuing role in the development of navigation facilities on the Bay. Since 1899 the federal government has regulated physical development on navigable waters and administered oil containment, nuisance weed removal, and debris removal programs. Most important to the water-quality story, the Public Health Service was an important source of early investigation on the Bay and later of supervising the shellfish sanitation program of the states. After World War II, it administered the programs established by the water pollution control acts until this function was transferred to the Department of the Interior in 1965. Since 1970 the United States Environmental Protection Agency has been in charge of federal efforts to improve water quality.

Still active are three research institutions on the Bay—Virginia Institute of Marine Science, Chesapeake Biological Laboratory, and Chesapeake Bay Institute. All have received a substantial amount of

The weather-beaten dock and old tree contrast dramatically with the modern towers of the Baltimore Gas and Electric Company's Wagner Power Plant. This fossil-fueled complex is located on Cox Creek, a tributary of the Patapsco River, just south of Baltimore. Courtesy Maryland Power Plant Siting Program, Maryland Department of Natural Resources.

their financial support from state and federal governments. The first two have been official state agencies for at least part of their histories. Their establishment, the first two around 1930, the latter in 1949, signaled the growing involvement of scientists in Bay governance.

Having surveyed the changing attitudes towards Chesapeake Bay over a 365-year history, the authors want to offer readers some perceptions of their own.

—Much of the present discussion of governance of the Bay proceeds on the basis of an idealized model of how Bay management operates. The ideal or model first assumes preparation of a scientific baseline study that establishes the present condition of Chesapeake Bay. Next comes the adoption of standards which choose the desired level of environmental quality for Bay waters. Then follows adoption of the comprehensive plan which selects the preferred uses of Chesapeake Bay from among those consistent with the quality standards. Finally comes an implementation strategy whereby government puts the plan into effect through public works and private regulation. Our study shows that the reality is much different.

—Chesapeake Bay may be the most studied estuary in the world, but scientists still have limited knowledge of its nature. Yet Bay scientists act as advocates and experts, bureaucrats and promoters. Although science so far has failed to provide a perfect baseline of information, scientists play a major role in shaping public perceptions of Bay problems.

—When making public choices affecting the quality of the Bay, government has responded more to public opinion and political pressure than to scientific analyses of its environmental and economic conditions.

Governments do not manage Chesapeake Bay, as is often said. The Bay as a natural resource is too complex, knowledge is often too limited, and public choices are frequently too fickle. Governments lack the capacity to "manage" nature.

Recognition that man has the capacity to alter the resources of the earth is not new. In 1863 George Marsh Perkins expressed this idea in his book *The Earth As Modified by Human Action.* Although this viewpoint did not come to occupy a central place in the history of science for another century, scientists from time to time have expressed concern for the Bay. In 1884 Dr. Charles W. Chancellor, Health Commissioner of Maryland, suggested, at least by implication, that the Bay could be ruined by the sewage from Baltimore.[6] In the 1930s the future of the Bay was discussed at the Chesapeake Bay Conference, and Dr. Truitt of the Chesapeake Biological Laboratory raised questions about the combined and cumulative effects of waste disposal.[7] What is new is the widespread public and political concern for the health of the Bay.

The ultimate question of whether man is ecologically endangering Chesapeake Bay lies beyond the scope of this book and the knowledge of its authors. Some observers argue that the Bay as a physically open system cannot be killed in any biological sense, and that man's perturbations are insignificant when compared to the changes wrought by natural phenomena like, for example, Tropical Storm Agnes in 1972.

But there is little question that technology has increased man's capacity to affect the Bay adversely. The kepone spill on the James River in 1975 has permanently poisoned its fishery; engineers and scientists recommend no clean-up. The 1979 malfunction of a nuclear power plant at Three Mile Island on the Susquehanna River, the Bay's major source of fresh water, environmentally threatened not only the whole Chesapeake Bay, but also the whole northeastern United States. These two events, both of which occurred after the study period ended, stand as stark witness to man's destructive capacity.

# Notes

CHAPTER 1

## FROM JAMESTOWN TO TROPICAL STORM AGNES

1. Carville Earle, 96.

2. Annie Dillard, *The Pilgrim of Tinker Creek* (New York: Harper's Magazine Press, 1974), 151.

3. Chesapeake Research Consortium, *The Effects of Tropical Storm Agnes on the Chesapeake Bay System*, 1976.

4. In 1964 the Maryland Attorney General ruled that a statute when referring to "rivers, creeks or branches" did not apply to bays: this opinion was subsequently overruled by the Maryland Court of Appeals. Board of Public Works v. Larmar Corp., 262 Maryland Reports, 24, and 277 Atlantic Reporter 20, (1971): 427.


CHAPTER 2

## A NOBLE ARM OF THE SEA

1. W. B. Cronin, *Volumetric, Areal and Tidal Statistics of the Chesapeake Bay Estuary and Its Tributaries* (Baltimore: Chesapeake Bay Institute, Special Report 20, 1971). Despite the definitive nature of this report, many Bay authors have continued to use other sources for Bay dimensions, thereby perpetuating old errors.

2. U. S. Army Corps of Engineers, *Chesapeake Bay: Existing Conditions Report*, Appendix C: Processes and Resources C-VI-2. This report is the result of multiple authorship spanning years of collection and compilation, actually contains several competing figures for various measurements for the Bay.

3. A. J. Lippson and R. L. Lippson, *The Condition of Chesapeake Bay and Assessment of Its Present State and Its Future.* Presented to the Marine Environmental Quality Committee (Warsaw, Poland: International Council for the Exploration of the Sea, Warsaw, Poland (1979), 1.

4. A particularly interesting discussion and analysis is contained in A. Y. Kuo, et al., *Chesapeake Bay: A Study of Present and Future Water Quality and Its Ecological Effects* (Gloucester Point, Va.: Virginia Institute of Marine Science, 1975), 1 ff.

5. U. S. Army Corps of Engineers, *Chesapeake Bay: Future Conditions Report*, Summary, 19.

6. *Ibid.*, 23.

7. Donald W. Pritchard, "Chemical and Physical Oceanography of the Bay" in State of Maryland, II, 49.

8. L. Eugene Cronin, "The Biology of the Chesapeake Bay," in State of Maryland, II, 77 ff.

CHAPTER 3

MANY DEEP WATER PORTS

1. William J. Hargis, Jr., *Exploration and Research in Chesapeake Bay: Being a Brief History of the Development of Knowledge of the Bay of Santa Maria* (Gloucester Point, Va.: Virginia Institute of Marine Science Contribution, No. 837 (1977), 14-48.

2. Richard Hakluyt, quoted in Reps, 24.

3. Carville Earle, 103.

4. Percy, lxii.

5. John Smith, quoted by Owens, 8-9.

6. The Virginia Council of the London Company, quoted in Reps, 31.

7. Reprinted in *Virginia Magazine of History and Biography*, 4:261.

8. Padover, 326.

9. Population figures come from Simmons, 171.

10. Boorstin, 106.

11. Starkey, 11.

12. Cox, xiv.

13. The Reverend Hugh Jones, *Present State of Virginia*, London, 1724, quoted in Chesapeake Research Consortium, 66.

14. Boorstin, 108.

15. Warren J. Coxe et al, *Guide to the Architecture of Washington, D. C.* (New York: Praeger, 1965), 3.

16. Coleman, 270.

17. Wertenbaker, quoted in "Annapolis: Intellectual Life around the Punch Bowl."

18. Everstine, *The Compact of 1785*.

19. Boorstin, 172; Billings, 323.

20. Randall Beirne, interview, December 1982.

21. Beirne interview.

22. John H. B. Latrobe used this phrase about men he knew in his youth, just after the turn of the nineteenth century—Samuel Smith, Robert Gilmor, Robert Oliver, and other merchant-princes.

23. Beirne interview.

CHAPTER 4

A WELL, A SINK, A PESTHOLE

1. John Smith, quoted in Burgess, 4.

2. Maryland Laws, Ch. 11 (1792).

3. Baltimore City Council, quoted in Scharf, *Chronicles*, 303.

4. Wertenbaker, *Norfolk*, 134.

5. Maryland Laws, Ch. 79 §13 (1808).

6. Maryland Laws, Ch. 6 (1886).

7. Maryland Laws, Ch. 355 (1874).

8. Report of the Baltimore Water Commissioner, (Baltimore, James Lucas, 1853), 142.

9. Baltmore v. Warren Manufacturing Co., 59 Maryland Reports, (1882), 84.

10. Commonwealth v. Webb, 6 Rand 27 Virginia Reports, (1828), 726.

11. Miller v. Truehart, 4 Leigh 31 Virginia Reports, (1833), 569.

12. 36 Southeastern Reporter (Virginia 1900), 373.

13. *Ibid.*

14. Charles Carroll the Barrister, letter to his agent William Anderson in London, September 27, 1762 in Michael Trostel, *Mount Clare* (Baltimore: Colonial Dames, 1982), 117.

15. Col. Landon Carter, "The Diary of Colonel Landon Carter, of Sabine Hallan the Northern Neck of the Rappohannock River," *William and Mary College Quarterly* 13: 159.

16. Buckler, 30-31. *See also* "The Almshouse at Calverton: The Beginning of Scientific Medicine," *Maryland State Medical Journal* (May 1966), 15: 84.

17. Chesney, 2:102-3.

18. Armstrong, *History of Yellow Fever in Norfolk.*

19. Maryland Laws, Ch. 93 (1801).

20. Maryland Laws, Ch. 200 (1874).

21. Maryland Laws, Ch. 12 §1 (1886).

22. Van Bibber, 16-21.

23. Charles Varle, *A Complete View of Baltimore with a Statistical Sketch, etc.* (Baltimore: S. Young, 1833), 11.

24. Buckler, *Baltimore:*, 10.

25. Buckler, *A History*, 40.

26. Howard, 83.

27. Wolman, *Water, Health and Society*, 348.

28. Olson, 54, 92.

29. Maryland Laws, Ch. 22 (1766).

30. Buckler, *Cholera*, 20.

31. Interview, August, 1980.

32. Boorstin, 238.

33. Thomas E. Bond, "Report of Baltimore Public Health Department, Baltimore, 1825," in *Baltimore City Health Department, the First Thirty-Five Annual Reports, 1815-1849* (Baltimore: Baltimore City Health Department, 1953).

34. Quoted in Reps, 216.

35. Leech, 77.

36. Buckler, *The Basin and Federal Hill*, 49.

37. Hall, 424.

38. Buckler, *The Basin and Federal Hill*, 19.

39. Baltimore City Council, June 1795.
40. Campbell, 257.
41. Baltimore City Council notation on manuscript in the Corner Collection, Maryland Historical Society, for February 1768.
42. Reps, 126.
43. Kanarek, 117.

CHAPTER 5

# TRADERS AND WATERMEN

1. Alsop, 450.
2. John Smith, quoted in Billings, 215.
3. George Percy, *Account of the Voyage to Virginia and the Colony's First Days*, quoted in Billings, 22-23.
4. Gregory, 226.
5. Power, 81-83.
6. Billings, 321.
7. Maryland Laws, Ch. 32 (1796).
8. Virginia Constitution, §175.
9. Everstine, *The Compact of 1785*. Under the 1785 Maryland-Virginia Compact, full property rights along the shores were guaranteed to citizens of both states, including fishing rights. All laws to preserve fish had to be made with the consent of both states.
10. Maryland Laws, Ch. 9 (1745).
11. Dorsey, 4.
12. 6 Hen. 69, 1748, Ch. 29, §§1-3.
13. Owners of mills were ordered to make slopes for the passage of fish by the following Virginia laws:
Rivanna and Hedgeman rivers: 8 Hen. 361, 1769 §81;
Meherrin River, 8 Hen. 583, 1772, Ch. 33;
Rapidan River, 9 Hen. 579, 1778, Ch. 39;
7 Hen. 321, 1759, Ch. 32. Amended 7 Hen. 590, 1762, Ch. 16 (for Appomattox River);
7 Hen. 423, 1761, Ch. 20;
7 Hen. 321, 1759, Ch. 32, and, amended to include the Appomattox River, 7 Hen. 590, 1762, Ch. 16.
14. Allan Howard's mill was torn down because it entirely obstructed the passage of fish (7 Hen. 423, 1761, Ch. 20).
15. Patuxent, Maryland Laws, Ch. 70 (1802); Monocacy, Maryland Laws, Ch. 79 (1806); Susquehanna, Maryland Laws, Ch. 91 (1813); Pocomoke, Maryland Laws, Ch. 70 (1862).
16. Council Papers of Virginia, 1698, *Virginia Magazine of History and Biography*, 21: 76.
17. Maryland Laws, Ch. 95 (1810).
18. "Letters from John F. Mercer to Richard Sprigg," Mercer Papers, April 19, 1797 in Pearson, 904-5.
19. Royall, 109-48 passim.

20. Ames, 185.

21. State Planning Commission, 3.

22. Phipps v. Maryland, 22 Maryland Reports (1864) 38.

23. McCready v. Virginia, 94 United States Reports (1876), 39.

24. Garrett Power, "More about Oysters Than You Wanted to Know," *Maryland Law Review* (1970), 30: 202-10.

25. Maryland Laws, Ch. 311 (1834).

26. Maryland Laws, Article 71 §15 (1860); also, Ch. 184 (1867) on who may take what, where, when, and how.

27. Virginia Laws, 3 Hen. 46, 47, 1691.

28. Virginia Laws, Ch. 36 §13 (1748). Penalties were periodically imposed, including one for casting dead bodies into rivers (3 Hen. 353, Ch. 27, 1705).

29. Quoted in Dulaney, 2.

30. Quoted in Reps, 222.

31. Virginia Laws, 2 Hen. 455, 1679 Ch. 31.

32. Virginia Laws, 2 Hen. 484, 1680 Ch. 15,

33. Virginia Laws, 4 Hen. 111, 1722 Ch. 7,

34. Virginia Laws, 4 Hen. 177, 1726 Ch. 7,

35. Virginia Laws, 4 Hen. 375, 1745 Ch. 18,

36. 4 Har. & McH., (1774), 540.

37. Maryland Laws, Ch. 5 (1768).

38. Maryland Laws, Ch. 24 (1783).

39. Maryland Laws, Ch. 45 (1791).

40. Maryland Laws, Ch. 45, §7 (1791).

41. Virginia legislators responded with the following laws which provided:

—for the clearing of the Appomattox and Pamunkey rivers, 6 Hen. 394, 1752, Ch. 40.

—for clearing the James an the Chickahominy rivers and extending navigation, 8 Hen. 148, 1765, Ch. 34.

—for clearing the Great Falls of the James and extending navigation. 8 Hen. 148, 1765, Ch. 34.

—for extending navigation on the Potowmack from Fort Cumberland to the Tidewater, 8 Hen. 570, 1772, Ch. 31.

—for improving navigation, incorporation of the James River Company, 11 Hen. 450, 1784, Ch. 19, and for other improvements, 11 Hen. 341, 1783, Ch. 25.

—opening and extending navigation of the Potomack and the Potomack Company incorporated, 11 Hen. 510, 1784, Ch. 43.

—opening, extending navigation on the Appomattox, 12 Hen. 591, 1787, Ch. 53, and for the same on the Chickahominy, 12 Hen. 382, 1786, Ch. 92.

42. Maryland Laws, Ch. 154, §18 (1817).

43. Maryland Laws, Ch. 75 (1814).

44. Maryland Laws, Ch. 35, §3 (1800).

45. Gordon Wolman.

46. Deric O'Bryan and Russell L. McAvoy, *Gunpowder Falls, Md.* (Washington, D. C.: Government Printing Office, 1966), 6.

47. Maryland Laws, Ch. 27 (1753).

48. Maryland Laws, Ch. 24 (1783).

49. Maryland Laws, Ch. 58(1872).

50. Olson, 55.

51. Lasson, 27.

52. Gordon Wolman, 26.

53. Garitee v. Baltimore, 53 Maryland Reports, (1880), 422.

54. Baltimore and Ohio Railroad v. Chase, 43 Maryland Reports (1875), 35.

55. Norfolk City v. Cooke, Virginia, 27 Grat (1836), 430.

56. Quoted in De Gast, The Lighthouses of the Chesapeake, 2.

CHAPTER 6

## LAND OF PLEASANT LIVING

1. Seth, 65-66.

2. Quoted in Robert and George Barrie, Cruises (Bryn Mawr, Pennsylvania: Franklin Press, 1909), 51-52.

3. Brewington, 221.

4. Wertenbaker, Norfolk, 294-95.

5. Brewington, 223.

6. Swepson Earle, 265.

7. Seth, 65-66.

8. Dr. John D. Godman, quoted in the Baltimore Sun, July 14, 1955.

9. Maryland Historical Society Magazine (1950), 201.

10. Cooke, "Impact of Pollution".

11. Olmsted Brothers, 103-6.

12. Quoted in Reps, 127, 70.

13. Reps, 75.

14. Reps, 4.

15. Seth, 64.

16. Brochure, Mount Vernon Ladies' Association of the Union.

17. Summerson, 542.

18. Alan Gowans, Images of American Living (New York: Harper & Row, 1972), 133.

19. Maryland Gazette, August 20, 1754, quoted in Scharf, The History of Maryland, 2: 13.

CHAPTER 7

## AN IMMENSE PROTEIN FACTORY

1. There is no single study of the Bay oyster industry, although there is an abundance of source materials. For an outline of oyster legislation and production figures from 1810 to 1910, see Cummings, Investigation of the Pollution of Tidal Waters, 18.

2. William Keith Brooks, The Oyster.

3. See, e.g., Governor Oden Bowie, Message to the Maryland General Assembly, 1870, House and Senate Documents, 1870, 18-21; Governor Frank

Brown, *Message to the General Assembly,* House and Senate Documents, 1894, 16.

4. Brooks, *Report of the Oyster Commission of 1884.*

5. For a detailed account of the industry *see* Charles Hirschfield, *Baltimore: 1870-1900—Studies in Social History* (Baltimore: The Johns Hopkins University Press, 1941), 42-62 *passim.*

6. State of Maryland, *Report of the Bureau of Labor and Statistics* (Annapolis, 1903) *passim.*

7. *See* Green, 22-24.

8. *See Report on the Fisheries and Waterfowl of Maryland,* House and Senate Documents, 1872, Document E, 37.

9. *See* news article, *Baltimore Sun,* Sept. 16, 1941.

10. Frye, *The Men All Singing.*

11. Chesapeake Bay Authority, 121.

12. News article, *Baltimore Evening Sun,* Jan. 14, 1939.

CHAPTER 8

## SEWAGE AND SHELLFISH

1. *Manual of American Water Works, 1888* (New York: Engineers News, 1889), passim.

2. Wolman and Geyer, *Report.*

3. For a particularly lively account of these issues in Baltimore, *see* Maryland State Board of Health, 1884, *Report,* 28-32, 140-78.

4. For an interesting summary of this process, *see* Martin V. Melosi, ed., *Pollution and Reform in American Cities, 1870-1930* (Austin, Texas: University of Texas Press 1980)

5. Crooks, 132-54.

6. City of Baltimore, Sewerage Commission, *Annual Report,* 1897..

7. Maryland State Board of Health, *Biennial Report 1886-1887,* 201.

8. City of Baltimore, Sewerage Commission, *Annual Report, 1897,* 55.

9. Speer, 35.

10. City of Baltimore, Sewerage Commission, *Annual Report, 1897,* 54.

11. *Ibid.,* 55.

12. *Ibid.,* 80.

13. Crooks, 132-54.

14. "Mayor's Message to the City Council of Baltimore for the Year 1897," Baltimore, 1898, 24, Maryland Room, Enoch Pratt Free Library, Baltimore, Md.

15. *Ibid.,* 25.

16. *Baltimore Sun,* Sept. 23, 1898.

17. City of Baltimore, Sewerage Commission, *Annual Report* 1899.

18. *Ibid.,* 15-16.

19. *Ibid.,* 17.

20. Maryland Laws. 349 (1904) (April 7, 1904).

21. City of Baltimore, Sewerage Commission, *Annual Report,* 1906, 17.

22. City of Baltimore, Sewerage Commission, *Report of the Board of Advisory Engineers and of the Chief Engineer on the Subject of Sewage Disposal*, 1906, 17.

23. *Ibid.*, 43.

24. Commissioners of Fisheries of Virginia, *Annual Report*, Fiscal Years 1910-1911, 11.

25. *Ibid.*, Fiscal Years 1913-1914, 14.

26. *Ibid.*, Fiscal Years 1915-1916, 12.

27. *Ibid.*, Fiscal Years 1914-1915, 12.

28. *Ibid.*, Fiscal Years, 1915-1916, 12.

29. *See* City of Hampton v. Watson, 89 Southeastern Reporter (Virginia 1916), 81-83.

30. *Ibid.*, 82.

31. *Ibid.*, 82.

32. Darling v. City of Newport News, 96 Southeastern Reporter (Virginia 1918), 307-15.

33. *Ibid.*, 308.

34. *Ibid.*, 309.

35. Darling v. City of Newport News, 249 U. S. Reports (1918), 540-44.

36. *Ibid.*, 542.

37. *Ibid.*, 543.

38. Commissioners of Fisheries of Virginia, *Annual Report*, Fiscal Years 1922-1923, 10.

39. Cummings, *Investigation of the Pollution of Tidal Waters.*

40. Cummings, *Investigation of the Pollution and Sanitary Conditions of the Potomac Basin.*

41. Speer, 24.

42. Office of Environmental Programs, Maryland Department of Health and Mental Hygiene, "Notes and Information Concerning the Oyster Industry 1925-1928," (bound file), Letter of October 29, 1925.

43. Maryland Department of Health *Annual Report*, 1934, 7.

44. Chesapeake Bay Authority, *Conference Report, Oct. 6, 1933*, 33.

45. Commissioners of Fisheries of Virginia *Annual Report*, Fiscal Year 1928-1929, 11.

46. Ibid., Fiscal Year 1930, 6.

47. Commonwealth v. City of Newport News, 164 Southeastern Reporter, (Virginia 1932), 690.

48. *Ibid.*, 692.

49. *Ibid.*, 689-700.

50. *Ibid.*, 700.

51. Virginia General Assembly. *Pollution: Report of the Committee Appointed by the Governor.* Senate Document No. 6, Division of Purchase and Printing, Richmond, Va., Jan., 1934, 3.

52. *Ibid.*, 8.

53. *Ibid.*, 9.

54. *Ibid.*, 17-32.

55. The purpose of the act was ". . . the relief of the district from pollution and the consequent improvement of conditions affecting the public health and the natural oyster beds, rocks, and shoals."

56. Commissioners of Fisheries of Virginia. *Annual Report*, Fiscal Years 1938-1939, 9.

57. *Ibid.*, 8.

CHAPTER 9

## DON'T LET THE FACTORIES IN

1. Virginia: *e.g.* Virginia Laws, 2 Hen. 484, 1680 Ch. 15. Maryland: *e.g.*, Maryland Laws, Ch. 79, §13 (1808).

2. Maryland Commissioners, 6-7.

3. Cummings, *Investigation of the Pollution of Tidal Waters*, 43.

4. Maryland Laws, Ch. 810 (1914), "The State Board of Health shall have general supervision and control over the waters of the States, insofar as their sanitary and physical condition affect the public health and comfort."

5. Maryland Laws, Ch. 14 (1917).

6. Maryland Conservation Department, *Annual Report*, 1922, 24,

7. *Ibid.*, 19.

8. Mark S. Watson, "The Fruits of Conservation," *Baltimore Sun*, July 13, 1930.

9. Department of Health of Maryland, *Annual Report*, 1936, 182.

10. Maryland Conservation Department, *Annual Report*, 1936, 14-15.

11. *Baltimore Sun*, Oct. 3, 1940.

12. Department of Health of Maryland, *Annual Report*, 1937, 195.

13. *Ibid.*, 197.

14. Quoted in Maryland Water Pollution Control Commission. *Water Pollution*, 10.

15. Maryland Conservation Department, *Annual Report*, 1940, 59.

16. *See* Water Pollution vertical file, Pratt Memorial Library, Baltimore, Md.

17. Interview, Dr. Corneillius Kruse, Department of Public Health, Johns Hopkins University, Baltimore, October 16, 1980.

18. *See* articles in the *Baltimore Sun*, dated May 10, 1936, August 28, 1938, and September 8, 1938.

19. *See*, e.g., Acts of the Virginia Assembly Ch. 61 (1849-1850); Ch. 147 (1852), Ch. 85 (1874); Ch. 270 (1884).

20. Levy, 21.

21. Maryland Conservation Department, *Annual Report*, 1922, 21.

22. Commissioners of Fisheries of Virginia, *Annual Report*, Fiscal Years 1918-1919, 10.

23. *See* discussions of oil pollution in Maryland Conservation Department, *Annual Report*, 1923, 27-31, and *Annual Report*, 1924.

24. Commissioners of Fisheries of Virginia, *Annual Report*, Fiscal Year 1930, 6.

25. Commissioners of Fisheries of Virginia, *Annual Report*, Fiscal Years 1938-1939, 30.

26. *Ibid.*, Fiscal Years 1940-1941, 12.

27. Pleasants, 108.

28. "Craney Island Spoil Disposal Area," briefing paper prepared by the Norfolk District, U. S. Army Corps of Engineers, June, 1970.

29. Chesapeake Bay Authority.

30. Huntsman.

31. *Ibid.*, 22.

32. *Ibid.*, 14.

33. *Ibid.*, 24.

CHAPTER 10

A BAY BUREAUCRACY

1. Commissioners of Fisheries of Virginia, *Annual Report*, Fiscal Years 1946-1947, 9.

2. *Ibid.*, Fiscal Years 1948-1949, 30.

3. This account of the origins of the State Water Control Board was given in a private interview by A. H. Paessler, long-time Executive Secretary to the Board, Richmond, Virginia, December 1980.

4. Commissioners of Fisheries of Virginia, *Annual Report*, Fiscal Years 1948-1949, 10.

5. Water Resources Policy Committee, 187.

6. A. H. Paessler interview. *See* note 3 above.

7. Commissioners of Fisheries of Virginia, *Annual Report*, Fiscal Years 1952-1953, 10.

8. *Ibid.*, Fiscal Years 1956-1957, 10.

9. *Ibid.*, Fiscal Years 1958-1959, 11.

10. *Ibid.*, Fiscal Years 1960-1961, 12.

11. *Ibid.*, Fiscal Years 1956-1957, 49.

12. *Ibid.*, Fiscal Years 1956-1957, 49-51.

13. *Ibid.*, Fiscal Years 1960-1961, 46.

14. *Ibid.*, Fiscal Years 1958-1959, 32.

15. *Ibid.*, Fiscal Years 1958-1959, 51-52.

16. Maryland Board of Natural Resources, *Annual Report*, 1945, 76.

17. *Ibid.*, *Report* 1947, 143.

18. *See Baltimore Sun* articles of November 27, 1945, March 16, 1946, and October , 1946 for discussions of the development of the act.

19. Maryland Annotated Code Article 66C, Sections 34-45, 1951.

20. Maryland Water Pollution Control Commission, *Annual Report*, 1947, 9.

21. *Ibid.*, *Biennial Report*, 1952-1953, 9.

22. *Ibid.*, *Biennial Report*, 1950-1951, 25.

23. Dr. Joseph McLain (first chairman of the Water Pollution Control Commission), private interview, Chestertown, Maryland, September 1980.

24. Department of Health of Maryland, *Annual Report*, 1949, 40-41.
25. *Baltimore Sun*, June 19, 1957.
26. Edgar L. Jones, "Maryland Pollution," *Baltimore Sun*, March 6, 1955.
27. *See Baltimore Sun*, March 12, 1955; and November 29, 1956.
28. Department of Health of Maryland, *Annual Report*, 1949, 49.
29. *See* Maryland Board of Natural Resources, *Annual Report*, 1947, 1959, and 1961.
30. *Baltimore Sun*, March 1, 1954.
31. *Washington Post*, November 1, 1957.
32. Wolman, et. al., *A Clean Potomac*, 49-50.

CHAPTER 11

SAVE THE BAY

1. A. H. Paessler, private interview, Richmond, Virginia, December 1980.
2. Reference to these Industrial Waste Symposiums appear in the Annual Reports of the Water Pollution Control Commission (after 1964 the Department of Water Resources) from 1961-1968.
3. Commissioners of Fisheries of Virginia, *Annual Report*, Fiscal Years 1960-1961, 12.
4. Virginia Marine Resources Commission (formerly Commissioners of Fisheries), *Annual Report*, Fiscal Years 1970-1971, 24.
5. *Ibid.*, Fiscal Year 1972, 23.
6. State of Maryland, Proceedings, II, 93.
7. *Baltimore Sun*, May 16, 1969.
8. *See* Annual Reports of the agency spanning this period.
9. Pleasants.
10. Nichols, 571f.
11. Virginia Marine Resources Commission, *Annual Report*, Fiscal Year 1972, 23.
12. Pleasants, 109.
13. Code of Virginia, 1950, section 28, 1-147.
14. Kanarek, 117.
15. Maryland Board of Natural Resources, *Annual Report*, 1961, Appendix B, 151 ff.
16. *Ibid.*, 153.
17. *Ibid.*, *Report*, 1966, 8-9.
18. Letter of Joseph H. Manning, Director of the Maryland Department of Chesapeake Bay Affairs, to Senator William James, April 16, 1971. Files of the Department of Natural Resources.
19. Maryland Department of Natural Resources Open File, "Submerged Lands Commission."
20. *Ibid.*
21. This was extracted from various files of the Maryland Department of Natural Resources. The letters cited are: James B. Coulter to Baltimore District, U. S. Corps of Enginers, dated November 17, 1970, and Joseph H. Manning to Paul McKee, dated August 27, 1971.

22. Maryland Department of Chesapeake Bay Affairs, Wetlands Hearing File, Hart-Miller Island Spoil Disposal Area.

23. Work on the Hart-Miller Island project was finally begun in 1982.

24. Natural Resources Institute, *"Patuxent Thermal Studies.*

25. Chesapeake at Bay "A Chilling View of the Thermal Threat," *Baltimore Sunpapers* Reprint, undated, unpaged.

26. Maryland Department of Natural Resources, Office of the Secretary, "Calvert Cliffs Nuclear Power Plant" File.

27. D. W. Pritchard, "An Appraisal of the Probable Effects of the Morgantown Electric Power Plant on the Estuarine Environment of the Potomac River," mimeo, undated (letter of transmittal to Senator Edward Hall, dated March 3, 1969), 1.

28. *Ibid.,* 8.

29. L. Eugene Cronin, "Letter to Donald W. Pritchard," mimeo, March 6, 1969, 3, 4.

30. *Ibid.,* 4.

31. *Ibid.,* 6.

32. William M. Eaton (Chairman), Nuclear Power Plants in Maryland, Report of the Governor's Task Force, Annapolis, Maryland, December 1969.

33. *Ibid.,* 51.

34. L. Eugene Cronin and Joseph A. Mihursky, press release, March 3, 1979. Mimeo, contained in "Calvert Cliffs" File, Office of the Secretary, Maryland Department of Natural Resources, Annapolis.

35. Eaton, Appendix C, 2.

36. *Ibid.,* Appendix D, 26.

37. *Baltimore News American,* August 28, 1976.

38. Baltimore Gas & Electric Company circular to customers, "Energy News," May 1981.

39. A. H. Paessler, private interview, Richmond, Virginia, December 1980.

40. Donald W. Pritchard, private interview, Ocean City, Maryland, October 1980.

41. *County Record* (Queen Anne's County, Maryland), February 26, 1969.

42. U. S. Congress, House Committee on Public Works, *The Chesapeake & Delaware Canal,* 91st Cong., 2d sess, 7, 8 April and 21 May 1970.

43. *Ibid.,* 113-25.

44. *Ibid.,* 273.

45. *Ibid.,* 299, *passim.*

46. Letter, Corps of Engineers Contracts Office to Governor Marvin Mandel, dated October 8, 1970. Contained in Submerged Land's Commission Vertical File, Department of Natural Resources Library, Annapolis.

47. Maryland Board of Natural Resources, *Report of the Water Resources Committee,: Annual Report,* 1961, 144 ff.

48. Maryland Laws, Chapter 82 (1964).

49. Federal Water Pollution Control Administration, *Chesapeake Bay: Susquehanna River Basins Project for Water Supply and Water Quality*

*Management,* (Washington, D. C.: Government Printing Office, 1965, Revised June 1966).

50. William Colony, private interview, Crystal City, Virginia, September 1980.

51. United States Statutes at Large, Section 312, Rivers & Harbors Act of 1956.

52. U. S. Army Corps of Engineers, Baltimore District, *Chesapeake Bay: Existing Conditions Report* and *Future Conditions Report.*

53. U. S. Department of the Interior, Federal Water Pollution Control Administration, *National Estuarine Pollution Study* (Washington, D. C.: Government Printing Office, 1969) vol. 3.

54. U. S. Department of the Interior, Fish & Wildlife Service, *National Estuary Study* (Washington, D. C.: Government Printing Office, 1970.), vol. 3.

55. *Ibid.,* 66.

56. *Remote Sensing of the Chesapeake Bay.*

57. State of Maryland. *Proceedings of the Governor's Conference.*

58. Wallace.

59. *See* Pleasants and Hargis.

60. *See* Abel Wolman, John C. Geyer, and E. E. Pyatt.

61. Maryland Annotated Code, Natural Resources Article, Title 8, Subtitle 12, 1974.

62. *Ibid.,* Subtitle 11.

63. Public Laws No. 92-500, Section 208, 33 U. S. C., Section 1288 (1973).

64. This is drawn from the personal experience of John Capper, who was Hearing Officer for approximately one hundred fifty Maryland wetlands cases in 1970-71.

65. Maryland Department of Natural Resources, Office of the Secretary, "Steuart Petroleum" File.

66. *Washington, D. C. Evening Star,* May 16, 1969.

67. The various dredging projects of the Harry Lundeberg School of Seamanship are contained in Maryland Department of Water Resources (now Water Resources Administration) waterway improvement files.

68. Maryland Annotated Code Natural Resources Article, Title 9, 1974.

69. Acts of the Virginia Assembly, Chapter 711, (1972).

CHAPTER 12

## PERCEPTIONS OF CHESAPEAKE BAY

1. Buckler, *A History of Epidemic Cholera,* 40.

2. *Baltimore Sun,* July 6, 1943. *See also* articles in the *Sun,* March 30, 1931, and April 20, 1950, and *Baltimore Evening Sun,* January 14, 1939.

3. *The Wall Street Journal,* April 24, 1953. These articles are contained in the Vertical Files of the Maryland Room, Enoch Pratt Free Library, Baltimore.

4. These articles are contained in the Vertical Files of the Maryland Room, Enoch Pratt Free Library, Baltimore.

5. Statement of Governor Marvin Mandel, before the Subcommittee of Intergovernmental Relations of the Senate Committee on Government Operation, February 4, 1970.

6. Maryland Department of Health, *Report 1884*, quoting House and Senate Document I, 1884, ". . . in the course of time, Chesapeake Bay will become a distillation of all the filth of Baltimore City," 178.

7. See Maryland Conservation Department, *Annual Report 1940*, 59 and Chesapeake Bay Authority, *Conference Report 1933*.

# Selected Bibliography

Alsop, George A. *A Character of the Province of Maryland.* London, 1666, vol. 1, edited by William Gowans. Baltimore: Maryland Historical Society Fund Publication, 1869.

Ames, Susie M. *Studies of the Virginia Eastern Shore in the Seventeenth Century.* New York: Russell and Russell, 1940.

Anderson, Alan D. *The Origin and Resolution of an Urban Crisis: Baltimore 1890-1930.* Baltimore: The Johns Hopkins University Press, 1977.

Armstrong, George D. *History of Yellow Fever in Norfolk.* Philadelphia, 1856.

Barbour, Philip. *The Jamestown Voyages under the First Charter, 1606-1609.* Cambridge, England: Cambridge University Press, 1969.

Beadenkopf, Anne. "The Baltimore Public Baths and Their Founder, the Rev. Thomas M. Beadenkopf," *Maryland Historical Magazine* (1950), 201-14.

Beers, Roland, et al. *The Chesapeake Bay.* Springfield, Va.: National Technical Information Service, 1971.

Billings, Warren M., ed. *The Old Dominion in the Seventeenth Century.* Chapel Hill: University of North Carolina Press, 1975.

Blair, Carvel Hall, and Ansel, Willits Dyer. *Chesapeake Bay: Notes and Sketches.* Cambridge, Md.: Tidewater Publishers, 1970.

Boorstin, Daniel J. *The Americans: The Colonial Experience.* New York: Random House, 1958.

Brewington, M. V. *Chesapeake Bay: A Pictorial Maritime History.* Cambridge, Md.: Cornell Maritime Press, 1976.

Brooks, William Keith. *The Development and Protection of the Oyster in Maine.* Baltimore: Johns Hopkins University Press, 1884.

———. *The Oyster.* Baltimore: Johns Hopkins University Press, 1891 (2d ed. 1905).

———. *Report of the Oyster Commission of 1884.* Baltimore: Johns Hopkins University Press, 1884.

Browne, Gary. *Baltimore in the Nation 1789-1861.* Chapel Hill: University of North Carolina Press, 1980.

Buckler, Thomas H. *Baltimore: Its Interest—Past, Present and Future.* Baltimore: Cushings & Bailey, 1878.

———. *The Basin and Federal Hill . . . Baltimore: Past Follies and Present Needs.* Baltimore, 1875.

———. *A History of Epidemic Cholera.* Baltimore: J. Lucas, 1851.

Burgess, Robert H. *This Was Chesapeake Bay.* Cambridge, Md.: Cornell Maritime Press, 1963.

Byrd, William, II. *Prose Works: Narratives of a Colonial Virginian,* edited by Louis B. Wright. Cambridge, Ma.: Harvard University Press, 1966.

Byron, Gilbert. *These Chesapeake Men.* North Montpelier, Vt.: Driftwood Press, 1942.

Campbell, Sir George, M. P. *White and Black.* New York, 1879.

Carter, Col. Landon. "Diary." *William and Mary College Quarterly,* 1904-1905.

Chesapeake Bay Authority. *Conference Report.* Baltimore, 1933.

Chesapeake Research Consortium. *The Effects of Tropical Storm Agnes on the Chesapeake Bay System.* Baltimore: Johns Hopkins University Press, 1976.

Chesney, Alan M. *The Johns Hopkins Hospital and the Johns Hopkins University School of Medicine: A Chronicle.* 2 vol. Baltimore: Johns Hopkins University Press, 1943.

City of Baltimore, Sewerage Commission. *Annual Report.* Baltimore, Fiscal Years 1897; 1899; and 1906.

———. *Report of the Board of Advisory Engineers and of the Chief Engineer on the Subject of Sewage Disposal.* Baltimore, 1906.

Coleman, R. V. *The First Frontier.* New York: Scribner's, 1948.

Commissioners of Fisheries of Virginia. *Annual Report.* Virginia, Fiscal Years 1910-1911; 1914-1915; 1915-1916; 1918-1919; 1922-1923; 1928-1929; 1930; 1938-1939; 1946-1947; 1948-1949; 1952-1953; and 1960-1961.

Cooke, Daniel. "Impact of Pollution on the Water-Oriented Activities of Back River, Maryland. Master's thesis, University of Tennessee, Knoxville, 1968.

Cox, Ethelyn. *Historic Alexandria, Virginia, Street by Street: A Survey of Existing Early Building.* Alexandria: Historic Alexandria Foundation, 1976.

Crooks, James B. *Politics and Progress: The Rise of Urban Progressivism in Baltimore 1895-1911.* Baton Rouge: Louisiana State University Press, 1968.

Cumming, Hugh S. *Investigation of the Pollution and Sanitary Conditions of the Potomac Basin.* Washington, D. C.: U. S. Public Health Service Hygienic Laboratory, Bulletin No. 104, 1916.

———. *Investigation of the Pollution of Tidal Waters of Maryland and Virginia, with Special Reference to Shellfish-Bearing Areas.* Washington, D. C.: U. S. Public Health Service Hygenic Laboratory, Bulletin No. 74, 1916.

Dabney, Virginius. *Richmond: The Story of a City.* New York: Doubleday, 1976.

De Gast, Robert. *The Bay.* Camden, Me.: International Marine Publishing Co., 1970.

———. *The Lighthouses of the Chesapeake.* Baltimore: Johns Hopkins University Press, 1973.

————. *Oystermen of the Chesapeake.* Camden, Me.: International Marine Publishing Co., 1970.

Dorsey, Emerson L., Jr. "A Legal History of the Port of Baltimore." Manuscript, University of Maryland School of Law, Baltimore, 1978.

Dulaney, Paul S. *The Architecture of Historic Richmond.* Charlottesville: University Press of Virginia, 1976.

Dowdy, Clifford. *The Golden Age: A Climate for Greatness, Virginia 1732-1775.* Boston: Little, Brown, 1970.

Earle, Carville. "Environment, Disease, and Mortality in Early Virginia," in *The Chesapeake in the Seventeenth Century: Essays in Anglo-American Society.* New York: Norton, 1979.

Earle, Swepson. *The Chesapeake Country.* Baltimore: Thomsen-Ellis Co., 1929.

Eaton, William M. *Nuclear Power Plants in Maryland, Report of the Governor's Task Force.* Annapolis, Md.: State of Maryland, 1969.

Embrey, Alvin T. *Waters of the State.* Richmond, Va: Old Dominion Press, 1931.

Evans, Ben. *Memories of Steamboating, Camp Meetings, Skipjacks and Islands of the Chesapeake.* Princess Anne, Md.: Marylander and Herald, Inc., 1977.

Everstine, Carl. *The Compact of 1785.* Annapolis, Md.: Legislative Council of Maryland, Research Report No. 26., 1946.

Footner, Hulbert. *Charles' Gift.* New York: Harper, 1939.

————. *Maryland Main and Eastern Shore.* New York: Appleton-Century, 1942.

————. *Rivers of the Eastern Shore.* Cambridge, Md.: Tidewater Publishers, 1979.

Frye, John. *The Men All Singing: The Story of Menhaden Fishing.* Norfolk: Dunning, 1978.

Garitee, Jerome Randolph. *Private Enterprise and Public Spirit: Baltimore Privateering in the War of 1812.* Ann Arbor, Mich.: University Microfilms, 1973.

Green, Harry J. *A Study of the Legislation of the State of Maryland.* Baltimore: Johns Hopkins University Press, 1930.

Greene, Suzanne Ellery. *Baltimore: An Illustrated History.* Woodland Hills, Ca.: Windsor Publications, Inc., 1980.

Gregory, William. "Journal from Fredericksburg, Virginia to Philadelphia." *William and Mary College Quarterly* 13 (1905).

Griffith, Thomas W. *Annals of Baltimore.* Baltimore: William Woody, 1824.

Guthein, Frederick. *Exploration and Research in Chesapeake Bay: Being a Brief History of the Development of Knowledge of the Bay of Santa Maria.* Gloucester Point: Virginia Institute of Marine Science, 1977.

————. *The Potomac.* New York: Rinehart & Co., 1968.

Hakluyt, Richard. *Discourse on Western Planning,* edited by Charles Deane. Maine: Documentary History of Maine, n.d.

Hall, Clayton Coleman. *Baltimore: Its History and Its People.* New York: Lewis Historical Publishing Co., 1912.

Hargis, William J., Jr. "James River Basin: Great Natural Resource or Problems of Developing the James River." Gloucester Point: Virginia Institute of Marine Science Mimeo., 1963.

Hirschfeld, Charles. *Baltimore 1870-1900*. Baltimore: Johns Hopkins University Press, 1941.

Howard, William Travis, Jr. *Public Health Administration and the Natural History of Disease in Baltimore, Md., 1797-1920*. Washington: The Carnegie Institution of Washington, 1924.

Huntsman, A. G. "Oceanographic Research on Chesapeake Bay." Undated. In the files of Maryland Department of Natural Resources, Annapolis. Typescript.

Jones, Edgar L. Jones. "Maryland Pollution," *The Baltimore Sunday Sun*, March 6, 1955.

Kanarek, Harold K. *The Mid-Atlantic Engineers: A History of the Baltimore District U. S. Army Corps of Engineers, 1774-1974*. Washington: Government Printing Office, 1977.

Land, Aubrey C. *Colonial Maryland*, Millwood, N. Y.: Kto Press, 1981.

Land, Aubrey C., Carr, Lois Green, and Papenfuse, Edward C. *Law, Society, and Politics in Early Maryland*. Baltimore: Johns Hopkins University Press, 1977.

Lang, Varley. *Follow the Water*. Winston-Salem: John F. Blair, Publisher, 1961.

Lasson, Kenneth. "A History of Potomac River Conflicts," in *Legal Rights in Potomac Waters*, edited by Garrett Power. Bethesda, Md.: Interstate Commission on the Potomac River Basin General Publication 76-2, 1976.

Leech, Margaret. *Reveille in Washington*. New York: Harpers, 1941.

Locke, Milo W. *Harbor of Baltimore: Locke's Plan*. Baltimore, 1875.

Levy, Ernest C. *Report to the Water Committee on the Investigation of the effect of trades wastes on the water of the James River at Richmond*. Richmond, 1905.

Lewis, Jack. *Potomac*. Dover, Del.: Woodwend Studios, n.d.

Mayer, Brantz. *Baltimore Past and Present*. Baltimore: Richardson & Bennett, 1871.

Maryland Board of Natural Resources. *Annual Reports*. Annapolis, Maryland: Fiscal Years 1945; 1947; 1959; and 1961.

Maryland Commissioners of Fisheries. *Annual Report*. Annapolis, Maryland: Fiscal Years 1902 and 1903.

Maryland Conservation Department. *Annual Report*. Annapolis, Maryland: Fiscal Years 1922; 1923; 1924; 1936; and 1940.

Maryland Department of Health. *Annual Report*. Annapolis, Maryland, Fiscal Years 1934; 1936; 1937; and 1949.

Maryland Department of Health and Mental Hygiene. *Notes and Information Concerning the Oyster Industry 1925-1928*. Annapolis, Maryland: Office of Environmental Programs, 1928.

Maryland State Board of Health. *Biennial Report*. Baltimore, 1886-1887.

Maryland State Board of Health. *Report*. Baltimore: House and Senate Document I, 1884.

Maryland Water Pollution Control Commission. *Annual Reports*. Annapolis Maryland, Fiscal Years 1947; 1950; 1951; 1952; and 1953.
————. *Water Pollution: A Policy and Program for Control*. Maryland, 1949.
Mencken, H. L. *Happy Days*. New York: Alfred A. Knopf. 1940.
————. *Newspaper Days*. New York: Alfred A. Knopf, 1941.
Middleton, Arthur Pierce. *Tobacco Coast: A Maritime History of Chesapeake Bay in the Colonial Era*. Newport News: The Mariners' Museum, 1953.
Miers, Earl Schenck. *The Drowned River*. Newark, De.: Curtis Paper Co., 1967.
Morgan, Edmund S. *Virginians at Home; Family Life in the Eighteenth Century*. Williamsburg, Va.: Colonial Williamsburg, 1952.
Morpeth, Lord. *Travels in America*. London, 1851.
*Mount Vernon*. Mount Vernon, Va.: Mount Vernon Ladies' Association of the Union, 1978.
Natural Resources Institute. *Patuxent Thermal Studies, Summary and Recommendations. Special Report No. 1*. College Park, Md.: Natural Resources Institute, 1969.
Nichols, M. M. "The Effect of Increasing Depth on the Salinity of the James River." Contribution No. 382, Geological Society of America. Gloucester Point: Virginia Institute of Marine Science. Mimeo.
Olmsted Brothers, *Report Upon the Development of Public Grounds for Greater Baltimore*, Baltimore, 1904.
Olson, Sherry. *Baltimore*. Baltimore: Johns Hopkins University Press, 1980.
Owens, Hamilton. *Baltimore on the Chesapeake*. Garden City, N. Y.: Doubleday, 1941.
Padover, Saul K., ed. *Thomas Jefferson and the National Capital 1783-1818*. Washington, D. C.: Smithsonian Institution Pamphlet.
Pearson, John C. *The Fish and Fisheries of Colonial North America*. Parts 4 and 5. Washington, D. C.: Fish and Wildlife Service, National Marine Fisheries Service, 1972.
Peden, William, ed. *Thomas Jefferson: Notes on the State of Virginia*. Chapel Hill: University of North Carolina, 1955.
Percy, George. *Observations Gathered Out of a Discourse of the Plantation of the Southerne Colonie in Virginia 1606*, London, 1625, reprinted in Edward Arber and A. G. Bradley, *Travels and Works of Captain John Smith*, vol. 1, lxii.
Pleasants, John B. *The Tidal James: A Review*. Special Report No. 18. Gloucester Point: Virginia Institute of Marine Science, 1971.
Power, Garrett, *Chesapeake Bay in Legal Perspective*, Washington, D. C.: Government Printing Office, 1970.
*Remote Sensing of the Chesapeake Bay: Conference Report*. Wallops Island, Va.: NASA, 1971.
Reps, John W. *Tidewater Towns: City Planning in Colonial Virginia and Maryland*. Chapel Hill: University of North Carolina Press, 1972.
Reynolds, Michael, et al., eds. *Maryland: A New Guide to the Old Line State*. Baltimore: Johns Hopkins University Press, 1976.

Ridgway, Whitman H. *Community Leadership in Maryland 1790-1840: A Comparative Analysis of Power in Society.* Chapel Hill: University of North Carolina Press, 1979.

Risjord, Norman K. *Chesapeake Politics 1781-1800.* New York: Columbia University Press, 1978.

Royall, Anne. *Sketches of the History, Life and Manners in the United States.* New Haven, Conn., 1826.

Scharf, J. Thomas. *Chronicles of Baltimore.* Baltimore: Turnbull Brothers, 1874.

———. *History of Baltimore City and County.* Philadelphia: Louis H. Everts, 1881.

———. *The History of Maryland.* Baltimore: Turnbull Brothers, 1879.

Semmes, Raphael. *Captains and Mariners of Early Maryland.* Baltimore: The Johns Hopkins University Press, 1931.

———. *Tidewater Boy.* Indianapolis: Bobbs-Merrill, 1952.

Seth, Joseph B. and Mary W. *Recollections of a Long Life on the Eastern Shore,* Easton, Md.: The Press of *The Star-Democrat,* 1926.

Shivers, Frank R., Jr. *Bolton Hill: Baltimore Classic.* Baltimore: Equitable Trust Co., 1978.

Simmons, R. C. *The American Colonies from Settlement to Independence.* New York: David MacKay, 1976.

*Somerset Iris and Messenger of Truth.* Princess Anne, Md., July 15, 1828.

Speer, Carl, Jr. "Sanitary Engineering Aspects of Shellfish Pollution." Master's thesis. Johns Hopkins University, Baltimore, 1936.

Starkey, Marion L. *Land Where Our Fathers Died: The Settling of the Eastern Shores.* New York: Doubleday, 1962.

State of Maryland. "Proceedings of the Governor's Conference on Chesapeake Bay." Unpublished typescript. Annapolis: Westinghouse Ocean Research and Engineering Center, 1968.

State Planning Commission. *Conservation Problems in Maryland.* Baltimore, 1935.

Summerson, Sir. John. *The Pelican History of Art: Architecture in Britain, 1530-1830.* Hamondsworth, Middlesex: Penguin, 1970.

Tate, Thad W. and David L. Ammerman, eds. *The Chesapeake in the Seventeenth Century.* Chapel Hill: University of North Carolina Press, 1979.

Tilp, Frederick. *This Was Potomac River.* Alexandria, Va., 1978.

Townsend, Richard. *Diary.* Unpublished diary in typescript. Enoch Pratt Free Library, Baltimore.

U. S. Army Corps of Engineers, Baltimore District. *Chesapeake Bay: Existing Conditions Report,* 7 Vols. Baltimore, 1973.

———. *Future Conditions Report,* 12 Vols. Undated.

U. S. Department of the Interior, Federal Water Pollution Control Administration. *Chesapeake Bay: Susquehanna River Basins Project for Water Supply and Water Quality Management.* Washington, D. C.: Government Printing Office, 1965 (Revised 1966).

———. *National Estuarine Pollution Study, vol. 3.* Washington, D. C.: Government Printing Office, 1969.

U. S. Department of the Interior, Fish and Wildlife Service. *National Estuary Study, vol. 3.* Washington, D. C., Government Printing Office, 1970.

Van Bibber, Dr. W. C. "The Drinking Waters in Maryland Considered with Reference to the Health of the Inhabitants." Baltimore: Medical and Chiurgical Society of Maryland, 1882.

Virginia Marine Resources Commission (formerly Commissioners of Fisheries). *Annual Report.* Fiscal Years 1970; 1971; and 1972.

Virginia State Senate. *Pollution: Report of the Committee Appointed by the Governor.* Richmond: Division of Purchase and Printing, 1934.

Wallace, McHarg, Roberts, and Todd, Inc. *Maryland Chesapeake Bay Study.* Philadelphia: Maryland Department of State Planning and the Chesapeake Bay Interagency Planning Committee, 1972.

Walsh, Richard and William Lloyd Fox. *Maryland: A History, 1632-1864.* Baltimore: Maryland Historical Society, 1974.

Warner, William W. *Beautiful Swimmers: Watermen, Crabs and the Chesapeake Bay.* Boston: Little, Brown, 1976.

Water Resources Policy Committee. *A Water Policy for the American People, Vol. I.* Washington, D. C.: Government Printing Office, 1950.

Watson, Mark S. "The Chesapeake Country's Life Revives," *Baltimore Sun,* July 13, 1930.

Wennersten, John R. *The Oyster Wars of Chesapeake Bay.* Centreville, Md.: Tidewater Publishers, 1981.

Wertenbaker, Thomas J. "Annapolis: Intellectual Life around the Punch Bowl," in *The Golden Age of Colonial Culture.* Annapolis, Md.: Department of Economic Development.

———. *Norfolk: Historic Southern Port* (2d ed.). Durham, N.C.: Duke University Press, 1962.

———. *The Shaping of Colonial Virginia.* New York: Russell and Russell, 1958.

Wilstach, Paul. *Tidewater Maryland.* Cambridge, Md.: Tidewater Publishers, 1969.

Wolman, Abel. *Water, Health and Society.* Bloomington: Indiana University Press, 1969.

Wolman, Abel and Geyer, John C. *Report on Sanitary Sewers and Waste Water Disposal in the Washington Metropolitan Region.* Baltimore: Johns Hopkins University, 1962.

Wolman, Abel, Geyer, John C., and Pratt, E. E. *A Clean Potomac River in the Washington Metropolitan Area.* Washington, D. C.: Interstate Commission on the Potomac River Basin, 1957.

# INDEX